中国野生动物

WILDLIFE IN CHINA

中国野生动物保护协会　编著
EDITED BY
China Wildlife Conservation Association

支持单位：北京 2022 年冬奥会和冬残奥会组织委员会
Supported by
Beijing Organising Committee for the 2022 Olympic and Paralympic Winter Games

中国林业出版社
China Forestry Publishing House

中国海关出版社
China Customs Press

自然影像中国

人与自然和谐共生

Harmonious Coexistence Between Man and Nature

大熊猫

拍摄地：陕西省

摄影：赵纳勋

Giant Panda

Ailuropoda melanoleuca

Location: Shaanxi Province

Photographer: Zhao Naxun

国家一级保护野生动物

早春雪后的秦岭，正在蹚过溪流的大熊猫。

大熊猫是中国野生动物保护的标志。中国建立了地跨四川、陕西、甘肃三省的大熊猫国家公园。大熊猫野外种群数量40年间从1114只增加到1864只。

State First-class Protected Wild Animal

A giant panda is wading through a stream in the Qinling Mountains after the snow in early spring.

The giant panda is a symbol of China's wildlife conservation. China has established the Giant Panda National Park across Sichuan, Shaanxi, and Gansu provinces. The wild population of giant panda has increased from 1,114 to 1,864 in 40 years.

东北虎

拍摄地：吉林省

摄影： 神鹰

Amur Tiger

Panthera tigris altaica

Location: Jilin Province

Photographer: Shen Ying

国家一级保护野生动物

吉林珲春森林中的东北虎。

虎是中国传统文化中的“百兽之王”，以虎为代表的大型食肉动物往往需要充足的猎物和大面积连片栖息地。2021 年 10 月，中国正式设立了东北虎豹国家公园，保护这一物种的野外种群及其生境。“百兽之王”正上演“王者归来”。

State First-class Protected Wild Animal

An Amur tiger in Hunchun Forest, Jilin Province.

The tiger represents the "King of Beasts" in traditional Chinese culture. Ample prey and habitats with high connectivity are the key elements to support the survival of large carnivores, like tiger. In October 2021, the Northeast Tiger and Leopard National Park was established to conserve the wild population and habitat for the return of the "King of Beasts".

川金丝猴

拍摄地：陕西省
摄影：丁宽亮
Golden Snub-nosed Monkey
Rhinopithecus roxellana
Location: Shaanxi Province
Photographer: Ding Kuanliang

国家一级保护野生动物

冬天的秦岭，一群川金丝猴挤在一起取暖。

川金丝猴是中国特有灵长类，种群数量 20000 多只，是中国金丝猴家族中最兴盛的种群。

State First-class Protected Wild Animal

A group of golden snub-nosed monkeys cuddled together against the cold winter of the Qinling Mountains.

The golden snub-nosed monkey, Chinese endemic primate, with a population of more than 20,000, is the most thriving population among all China's snub-nosed monkey species.

雪豹

拍摄地：青海省
摄影：奚志农
Snow Leopard
Panthera uncia
Location: Qinghai Province
Photographer: Xi Zhinong

国家一级保护野生动物

青海祁连山，一只雪豹在岩洞中向外张望。

雪豹主要分布在青藏高原和中亚山地，海拔 2000~5000 米。目前中国拥有全球雪豹一半以上种群数量和 60% 的栖息地。雪豹是青藏高原和周边山地生态系统中的伞护种，对雪豹的保护可以惠及整个生态系统。

State First-class Protected Wild Animal

A snow leopard looks out from a cave in Qilian Mountain, Qinghai Province.

Snow leopards mainly inhabit in the Qinghai-Tibet Plateau and the mountains of Central Asia, with an altitude of 2,000 to 5,000 meters. More than half of the global population and 60 percent of the habitat of snow leopard locate in China. As an umbrella species, conservation of the snow leopards can benefit the entire ecosystem of the Qinghai-Tibetan Plateau and surrounding mountains.

亚洲象

拍摄地：云南省

供图：中国野生动物保护协会

Asian Elephant

Elephas maximus

Location: Yunnan Province

Provider: China Wildlife Conservation Association

国家一级保护野生动物

2021 年 5 月，一群北移的亚洲象在树林中躺地休息。

亚洲象在中国主要分布在云南省西南边境地区，西双版纳是最集中分布的区域。种群数量从 1985 年的约 180 头增加到现在的 300 多头。

State First-class Protected Wild Animal

In May 2021, a herd of Asian elephants moving northward lay down in the woods for resting.

The Asian elephant populations in China mainly distribute in the southwest border area of Yunnan Province, especially concentrated in Xishuangbanna. The population in China has increased from roughly 180 in 1985 to over 300 in 2021.

朱鹮

拍摄地：陕西省
摄影：胡琳
Crested Ibis
Nipponia nippon
Location: Shaanxi Province
Photographer: Hu Lin

国家一级保护野生动物

秦岭山区柿树上停栖的朱鹮。

朱鹮在 1981 年被重新发现时仅剩 7 只个体。依靠严格的栖息地保护及人工繁育技术突破，2021 年，这个曾经几近灭绝的物种，野外种群和人工繁育种群总数在中国已有 5000 余只，为世界创造了拯救濒危物种的成功范例。

State First-class Protected Wild Animal

Crested ibises perch on persimmon trees in the Qinling Mountains.

Only 7 individuals remained, when this species was rediscovered in 1981. Relying on strict habitat protection and technical breakthroughs of captive breeding, the total number of wild and captive breeding populations of the species which almost extinct in China has reached more than 5,000 individuals, as a successful example of endangered species restoration.

大天鹅

拍摄地：北京市

摄影：谢建国

Whooper Swan

Cygnus cygnus

Location: Beijing

Photographer: Xie Jianguo

国家二级保护野生动物

海坨山下，一群大天鹅觅食归来，隐约可见北京2022年冬奥会和冬残奥会的高山滑雪赛场。邻近的北京延庆野鸭湖湿地，每年冬季大天鹅等候鸟在此栖息越冬。

State Second-class Protected Wild Animal

A flock of whooper swans came back from foraging at the foothill of Haituo Mountain, where the alpine skiing field of Olympic and Paralympic Winter Games Beijing 2022 locates. The nearby Yeyahu wetland in Yanqing, Beijing was the overwintering habitat of many migratory birds every winter, like whooper swans.

小天鹅

拍摄地：北京市
摄影：徐永春
Tundra Swan
Cygnus columbianus
Location: Beijing
Photographer: Xu Yongchun

国家二级保护野生动物

迁徙季节，北京野鸭湖血色夕阳里的小天鹅一家在新建的官厅水库特大桥下的水域栖息。

北京至张家口的京张高铁官厅水库特大桥跨越官厅水库，是北京 2022 年冬奥会和冬残奥会重点配套的基础设施，对于京津冀一体化发展具有重要意义。

State Second-class Protected Wild Animal

A family of tundra swans perche in the waters under the newly built Guanting reservoir bridge in the sunset of Yeyahu wetland, Beijing during the migration season.

The Guanting reservoir bridge of the Beijing-Zhangjiakou high-speed railway is a key supporting infrastructure for the Olympic and Paralympic Winter Games Beijing 2022, which plays a significant role in the collaborative development of the Beijing-Tianjin-Hebei Province region.

▮ 普通鵟

拍摄地：北京市

摄影：徐永春

Eastern Buzzard

Buteo japonicus

Location: Beijing

Photographer: Xu Yongchun

国家二级保护野生动物

北京的西山位于太行山余脉，南北走向的太行山与其东边的华北平原形成气流的上扬。每年春秋季节，大群猛禽选择这条路线迁徙，在喧闹的城市上空飞过。

State Second-class Protected Wild Animal

The Western Hill of Beijing locates in the branch of Taihang Mountain. The north-south Taihang Mountain and the North China Plain to the east form an upward air current. Every spring and autumn, large flocks of raptors choose this migratory route to fly over the noisy city.

灰鹤

拍摄地：河北省
摄影：谢建国
Eurasian Crane
Grus grus
Location: Hebei Province
Photographer: Xie Jianguo

国家二级保护野生动物

位于河北怀来的官厅水库湿地是天鹅、灰鹤、灰雁等候鸟的栖息地。每年冬季，数千只灰鹤在此栖息越冬。白天，它们飞到田野觅食，傍晚回到结冰的湖面过夜。

State Second-class Protected Wild Animal

The Guanting Reservoir Wetland in Huailai, Hebei Province is the habitat of migratory birds like swans, Eurasian cranes and graylag geese. Thousands of Eurasian cranes overwinter here. They forage in the fields during daytime and return to the frozen lake in the evening.

太平鸟

拍摄地：北京市
摄影：谢建国
Bohemian Waxwing
Bombycilla garrulus
Location: Beijing
Photographer: Xie Jianguo

冬季，临近首钢园区的北京冬奥公园，一群追觅浆果的太平鸟站在枝头，串串银铃般的叫声打动人们，勾勒出绿水青山门头沟的生动画卷。

In the Beijing Winter Olympic Park, near the Shougang Park, a flock of Bohemian waxwings hunting for berries stand on branches in winter. Their figures and touching silver bell-like singing paint a picture of beautiful scene of lucid waters and lush mountains of Mentougou.

序

野生动物是自然生态系统中最活跃、最引人注目的组成部分，是全人类的共同财富，承载着人类的精神文化价值，关乎地球生态安全。中国是世界上野生动物资源最为丰富的国家之一，据统计，中国仅脊椎动物就达7300种，占全球种类总数的10%以上。物种多样性高、特有种多、区系起源古老是中国野生动物资源的重要特征。

中华民族历来强调天人合一、尊重自然，守护大自然留给中国宝贵的野生动物资源。中国政府通过不断完善野生动植物保护法律法规体系、有效履行野生动植物保护行政管理和执法监督、打击野生动植物非法贸易、普及和提高公民的环境保护意识、加强和拓展双边及多边国际合作,建立了行之有效的综合管理体系,形成了中国特色的野生动物保护管理模式。

中国野生动物保护事业持续健康的发展，得益于中国政府对生态建设的高度重视，得益于社会公众对生态保护的大力支持。中国政府前所未有的重视生态文明建设和生态环境保护，野生动物保护和管理得到了极大的发展,野生动物种群数量得到恢复,栖息地质量得到改善。

作为一个负责任的大国，中国在致力于做好保护本国野生动物资源、实现野生动物资源可持续利用的同时，主动参与野生动物保护的国际事务，分享中国成功经验，认真履行所承担的国际义务，共同推动全球野生动物保护事业的可持续发展。

第24届冬季奥运会将于2022年2月4日至20日在北京和张家口举行，吉祥物“冰墩墩”就是以大熊猫为原型进行设计创作的，体现了人与自然和谐共生的理念。中国野生动物保护协会、飞羽视界文化传媒结合我国野生动物的生态习性及北京冬奥会和冬残奥会可持续发展的理念，编辑出版了《中国野生动物》画册，我真诚地希望通过精美的图片，让公众更多地了解中国多姿多彩的野生动物，感受生命的力量和自然野性之美。

是为序。

中国科学院院士　魏辅文

PREFACE

Wildlife is the most active and attractive components of nature and is the common fortune of all humankind. It is fortunate to share the world with wildlife, for its spiritual and cultural value, as well as wildlife's role in global ecosystems. China is among the countries with the most abundant wildlife resources. Over 7,300 species of vertebrates are recorded for China—around 10% of the world's wildlife. The wildlife resources in China feature high numbers of endemic species, many species with ancient origin, and high species richness across the country.

Chinese people highly value and respect the harmony between people and the rest of nature, and guard theses precious resources. The Chinese approach has been improving the legal system, ensuring effective implementation and supervision and tackling illegal trading, through promotion of bilateral and multi-lateral cooperation. Increased public awareness of environmental protection and establishment of an effective and comprehensive management system mean wildlife conservation in China is strong.

The consistent development of China's wildlife conservation approach has benefited from the Chinese government's focus on developing an ecological civilization and the dedicated support of the public for ecological conservation. China's government has paid unprecedented attention to eco-civilization, environment protection, and nature conservation. Through this focus, wildlife conservation and management has been improved, wildlife populations has continuously increased and habitat quality has been recovered.

Besides, as a responsible country, China is actively engaged in international activities of wildlife conservation, sharing successful experiences, and fulfilling international obligations, to improve the global wildlife conservation.

The 24th Olympic Winter Games will be held from 4th to 20th February 2022 in Beijing and Zhangjiakou. Its mascot "Bing Dwen Dwen" is a giant panda, as a symbol of harmony of people and nature. Combing the ecological habits of wildlife and the concept of sustainable development of the Olympic and Paralympic Winter Games Beijing 2022, *Wildlife in China* is published by China Wildlife Conservation Association and Bird Media Ltd. With these superb photographs, I hope that these colorful and wonderful wildlife will be known by more people, and that the beauty and power of nature will reach the public.

Academician of Chinese Academy of Sciences   Wei Fuwen

亚洲象

拍摄地：云南省
摄影：武明录
Asian Elephant
Elephas maximus
Location: Yunnan Province
Photographer: Wu Minglu

国家一级保护野生动物

2021年5月—8月，15头北上及返回的亚洲象之旅，引起国内外的广泛关注，让人们看到了中国保护野生动物的成果。云南正积极推进亚洲象国家公园建设，对亚洲象栖息地进行整体化的保护。

State First-class Protected Wild Animal

From May to August 2021, the northbound and return journey of 15 Asian elephants in Yunnan Province, which aroused widespread attention at home and abroad, is the vivid results of China's wildlife conservation. Yunnan Province is actively promoting the establishment of the Asian Elephant National Park for comprehensive habitat conservation.

前言

随着我国生态环境的改善，野生动物对于公众不再陌生。看野生动物，实际上不需要大费周折，因为我们就生活在一个野生动物的王国里，只要你留心。

来到公园，或到野外，闭上双眼，最能挑动你神经的是鸟儿清脆的鸣声；睁大双眼，最能激发你热情的是林影间跳跃的精灵。没错！这就是丰富多彩的野生动物王国。

影像是人类认识和了解野生动物的重要途径，用影像的力量唤起公众对野生动物保护的关注，已成为树立尊重自然、顺应自然、保护自然的生态文明理念的一种有效方式。得益于国家对生态文明建设的重视，中国的野生动物资源正在逐步恢复，越来越多的野生动物出现在我们的周边。希望越来越多的人走进田野和山川，用镜头去捕捉野生动物的精彩瞬间，用爱心来呵护我们赖以生存的绿水青山。

今天，奥林匹克运动“更快、更高、更强——更团结”的格言已经远远超出体育竞技的范畴，成为全人类的文明遗产，其丰富内涵和对于人类社会发展的重要性正在与日俱增。北京2022年冬奥会将于2022年2月4日至20日在北京和张家口举行，绿色办奥不仅是中国对庄严承诺的落实，也是向世界展示建设美丽中国的决心和信心。《中国野生动物》是我们献给冬奥会的一份礼物，希望能为世界各地的运动员和嘉宾朋友打开一扇了解中国生物多样性的大门。

结合我国野生动物的生态习性及北京2022年冬奥会和冬残奥会可持续发展的理念，我们选取了一部分有代表性的野生动物的照片，从不同视角展现了中国这片神奇、美丽的土地上，能体现“更快、更高、更强——更团结”的精彩瞬间，呈现了我国闻名世界的野生动物的多样性，体现了中国自然保护的卓越成就。我们希望通过这些精美的图片，公众不仅能够领略到作者在作品中蕴藏的爱心和情感，更能感受这些作品带来的心灵深处的震撼和美的享受。我们的目标是，用摄影语言讲出精彩的中国野生动物保护故事，用影像的力量推动全球的生物多样性保护。

野生动物真正的家在野外，保护它们的最好方式是维护其自然家园的完整性和原真性，让野生动物自由自在的生活，这是野性的呼唤，也是人性的呼唤。

编委会

2021年12月

FOREWORD

With improvements in the natural environment, wildlife is no longer strange to the public. If you keep your eyes open, not much effort is needed to watch them, as we live in a world surrounded by wildlife.

In the parks or the wild, close your eyes, the clear songs of birds inspire you; open your eyes, the animals among forests amaze you. That is the fantastic wildlife kingdom.

A key approach for people to get knowledge of wildlife, is through powerful images to get attention from the public. This is an effective way to instill in the public the eco-civilization concept of respecting nature, knowing and understanding nature, and conserving nature. Due to the emphasis of ecological civilization, wildlife resources are undergoing a gradual recovery, and more and more wildlife is to be seen. We hope more people to spend in the fields and the wild, to capture the wonderful moments of wildlife with cameras, and to care for clear rivers and green mountains on which we depend.

The motto of Olympics "Faster, Higher, Stronger—Together" is beyond simply sports competition. As a civilization heritage of all humankind, its rich meaning and importance to the development of society are consistently increasing. The 24th Winter Olympic Games will be held from 4th to 20th February 2022 in Beijing and Zhangjiakou, Hebei Province. Green Olympic is a fulfilling of commitment as well as resolution and confidence to construct "beautiful China". Dedicated to the Winter Olympics, we hope the book *Wildlife in China* could open a door for all involved in the games to know better the wildlife of China.

Combing the ecological habits of wildlife and the concept of sustainable development of the Olympic and Paralympic Winter Games Beijing 2022, some representative photographs from varied perspectives were collected to show wildlife-capturing moments of "Faster, Higher, Stronger—Together", and the achievements of nature conservation. Through these beautiful photographs, the empathy of each photographer can easily be felt. In this book, we aim to tell the story of wildlife conservation in China with photos and improve the global wildlife conservation through these images.

Habitats or the nature, are the home of wildlife, so the best way for wildlife conservation is to keep the integrity and originality of their habitats. To ensure the wild animals stay free is the call of the wild, as well as the call of the humanity.

Editorial Board
December, 2021

滇金丝猴

拍摄地：云南省
摄影：奚志农
Yunnan Snub-nosed Monkey
Rhinopithecus bieti
Location: Yunnan Province
Photographer: Xi Zhinong

国家一级保护野生动物

摄于1995年的这幅《母与子》照片，被多家国内外期刊转载。照片中滇金丝猴与人类别无二致的神态，透过镜头直达人的内心，使越来越多的人了解了这一物种。

State First-class Protected Wild Animal

Taken in 1995, this photograph of "Mother and Son" has been reprinted by many domestic and foreign journals. The uncanny resemblance between Yunnan snub-nosed monkeys and human has attracted more and more people to know about this species.

目录 CONTENTS

概述

在自然界中生活的野生动物充满着活力和野性，既有善于奔跑和跳跃的运动健将，也有经常为争夺食物、配偶而搏斗的挑战者。这些真实的场景，最大限度地呈现出野生动物的自然美、野性美、生命美。

中国幅员辽阔，地形复杂，气候多样，演化出丰富的植被类型和生境多样性，孕育了多种多样的野生动物。中国是野生动物资源最丰富的国家，也是世界生物多样性特别丰富的国家之一。据统计，中国仅脊椎动物就达7300种，占全球种类总数的10%以上；中国的野生动物区系起源古老，珍稀物种丰富，特有属、特有种多，是大熊猫（*Ailuropoda melanoleuca*）等400多种中国特有野生动物的家园。

北京，中国的首都，地形地貌复杂，有森林、灌丛、草甸、湿地等多种生态系统，生境类型多样，是世界上野生动物最为丰富的大都市之一。北京鸟类有500多种，物种之丰在世界各大都市中排名第二。通过再引入的方式，麋鹿(*Elaphurus davidianus*)从英国的乌邦寺重新回到北京的南苑故里，已成为野生动物保护的成功案例。毗邻冬奥会比赛场地的松山、野鸭湖以及海坨山，既是中国著名的自然保护区，也是野生动物的重要栖息地，豹猫（*Prionailurus bengalensis*）、中华斑羚（*Capricornis milneedwardsii*）、金雕（*Aquila chrysaetos*）、勺鸡（*Pucrasia macrolopha*）、灰鹤（*Grus grus*）、大天鹅（*Cygnus cygnus*）等动物经常出没。北京冬季温度低、降水少、风速大，冬季还常迎来皑皑白雪。虽然气候寒冷，但北京冬天依然可以看到很多野生动物。在郊外山区，灰鹤、黑鹳（*Ciconia nigra*）、褐马鸡（*Crossoptilon mantchuricum*）、岩松鼠（*Sciurotamias davidianus*）、狍（*Capreolus pygargus*）等成群活动；在城市公园，绿头鸭（*Anas platyrhynchos*）、鸳鸯（*Aix galericulata*）、长耳鸮（*Asio otus*）、太平鸟（*Bombycilla garrulus*）、红嘴蓝鹊（*Urocissa erythrorhyncha*）等与市民朝夕相伴。

在中国，野生动物已经深深融入中华民族的文化艺术之中。许多野生动物被作为图腾，虎、鹰等强壮威猛的动物，歌声委婉的鸣禽，美丽长寿的鹤类和龟类等，长期以来出现在戏曲、电影、诗篇、美术、书法、摄影、绘画、手工艺品以及建筑装饰中。野生动物所蕴含的丰富的文化、艺术内涵以及人与动物之间特殊的情感交流，潜移默化地影响着公众的价值观、行为和生活方式。

中国政府高度重视野生动物保护工作，不断完善以《中华人民共和国野生动物保护法》为核心的野生动物保护法律体系和管理制度，积极推进以国家公园为主体的自然保护地体系建设，大力实施野生动物迁地保护和回归自然，有效保护了90%的陆地生态系统类型、65%的高等植物群落、71%的国家重点保护野生动植物。大熊猫、朱鹮（*Nipponia nippon*）、黑颈鹤（*Grus nigricollis*）、藏羚（*Pantholops hodgsonii*）等珍稀濒危野生动物种群实现恢复性增长。

在那片茂密的森林、风景如画的湿地及一眼望去荒凉的高原大地中，野性之美，就在你的面前。

INTRODUCTION

Animals in nature are wild and free, running and jumping, as well as fighting for food and mates. These real scenes of wild animals extensively exhibit the natural beauty, wildness and life.

As one of the world's largest countries, China has complex terrain and diverse climates. Diverse vegetation types and habitats have evolved over time, contributing to high diversity of wild animals. China is amongst the countries with highest biodiversity including wild animal resources. Over 7300 species of vertebrates have been recorded, amounting to 10% of the world's species, many of ancient origin, with over four hundred endemic species including giant panda (*Ailuropoda melanoleuca*).

Beijing is the capital city of China, but also has a complex of terrain and diverse ecosystems, e.g., forests, shrublands, grasslands, and wetlands. Amongst the big cities in the world, Beijing has the highest diversity of wild animals. Beijing is also ranked the second highest in bird species diversity in cities of China, with over five hundred species. The Pere David's deer (*Elaphurus davidianus*) was re-introduced to Nanyuan of Beijing from Woburn Abbey, Britain, which is a successful case of captive breeding and reintroduction leading to wildlife conservation. Song Mountain, Wild Duck Lake, and Haituo Mountain nature reserves adjacent to the venues of Winter Olympics, are all important habitats for wild animals, e.g., leopard cat (*Prionailurus bengalensis*), Chinese goral (*Capricornis milneedwardsii*), golden eagle (*Aquila chrysaetos*), koklass pheasant (*Pucrasia macrolopha*), Eurasian crane (*Grus grus*), whooper swan (*Cygnus cygnus*). With low temperature, little precipitation and high wind speed in winter, and snowing is usual in Beijing. While wild animals are not limited by the cold climate and can be seen during winter. In the hilly areas of suburban Beijing, Eurasian crane, black stork (*Ciconia nigra*), brown-eared pheasant (*Crossoptilon mantchuricum*), Chinese rock squirrel (*Sciurotamias davidianus*), roe deer (*Capreolus pygargus*) *etc*. occur in groups; in the urban parks, mallard (*Anas platyrhynchos*), Mandarin duck (*Aix galericulata*), long-eared owl (*Asio otus*), bohemian waxwing (*Bombycilla garrulus*), *etc*. are constant companions to citizens.

Wild animals are part of the culture and art in China. Many animals are cultural symbols in opera, film, poetry, painting, calligraphy, photography, handicrafts, and decoration of architecture, e.g., strong and powerful tiger and eagle; songbirds with pleasant and musical tones; graceful and long-lived cranes and turtles, *etc*. The rich cultural and artistic links that wild animals have and the special emotional exchange between people and animals have influenced the values, behaviors and lifeways of citizens.

Chinese government attaches great importance to wildlife conservation: legal systems and management regulations based on the *Law of the People's Republic of China on Wildlife Conservation* are undergoing improvement; a system for natural protected areas emphasizing national parks is being actively promoted; *ex-situ* conservation of wild animals and rehabilitation and release back to the wild is a key management feature; 90% of terrestrial ecosystems, 65% of higher plant communities, and 71% of State key protected wildlife are well conserved. Population size of some rare and endangered wild animals have recovered and increased, e.g., giant panda, crested ibis (*Nipponia nippon*), black-necked crane (*Grus nigricollis*), Tibetan antelopes (*Pantholops hodgsonii*).

Among the dense forests, in scenic wetlands, and on the desolate plateau, the beauty of wildlife is in all around!

大熊猫
拍摄地：四川省
供图：卧龙国家级自然保护区
Giant Panda
Ailuropoda melanoleuca
Location: Sichuan Province
Provider: Wolong National Nature Reserve

国家一级保护野生动物

中国大熊猫保护研究中心核桃坪基地，正在进行野化训练的大熊猫母子。

State First-class Protected Wild Animal

Female giant panda and its cub are undergoing rewilding training at the Hetaoping Base of the China Conservation and Research Centre for the Giant Panda.

哺乳动物

MAMMALS

哺乳动物是动物界中最高等的一个类群，是最成功的适应者。哺乳动物是生态系统中不可或缺的重要成员，维系着各种生态系统的健康与稳定，很多物种如大熊猫（*Ailuropoda melanoleuca*）、东北虎（*Panthera tigris altaica*）、雪豹（*Panthera uncia*）成为生态系统健康的指示物种、旗舰物种和伞护物种。

中国已知野生哺乳动物686种，是全球哺乳动物物种多样性最为丰富的国家之一，约占世界兽类总数的10%。珍稀、濒危和国家重点保护野生哺乳动物，如大熊猫、金丝猴（*Rhinopithecus* spp.）、东北虎、雪豹、亚洲象（*Elephas maximus*）、中华穿山甲（*Manis pentadactyla*）等物种是世界范围备受关注的重点保护野生动物。

Mammals are the highest class in the animal kingdom, and the highly adaptive to all climates and habitats. Mammals maintain the health and stability of ecosystems. Many mammal species such as giant panda (*Ailuropoda melanoleuca*), Amur tiger (*Panthera tigris altaica*), snow leopard (*Panthera uncia*) are indicators of healthy ecosystems and are used as flagship species and umbrella species.

686 species of wild mammals have been recorded in China, making it amongst the countries with highest mammal diversity and accounting for about 10% of the mammal species of the world. Rare, endangered or nationally protected wild mammals are of significant conservation concern globally e.g., giant panda, snub-nosed monkey (*Rhinopithecus* spp.), Amur tiger, snow leopard, Asian elephant (*Elephas maximus*), Chinese pangolin (*Manis pentadactyla*).

大熊猫

拍摄地：陕西省
摄影：雍严格
Giant Panda
Ailuropoda melanoleuca
Location: Shaanxi Province
Photographer: Yong Yange

国家一级保护野生动物

陕西佛坪国家级自然保护区内，野生大熊猫哺育幼仔的难得瞬间。

State First-class Protected Wild Animal

A rare moment of wild giant panda feeding its cub in the Foping National Nature Reserve, Shaanxi Province.

亚洲象

拍摄地：云南省
摄影：谢建国
Asian Elephant
Elephas maximus
Location: Yunnan Province
Photographer: Xie Jianguo

国家一级保护野生动物

一群野生亚洲象进入野象谷景区，景区提前预警，游人安全避让。随着亚洲象种群的不断扩大，如何解决人象冲突成为亚洲象保护的新课题。

State First-class Protected Wild Animal

A herd of wild Asian elephants enter wild Elephant Valley Scenic Area, for which early avoidance warning has been given to tourists. With the growing of the Asian elephant population, how to resolve the human-elephants conflict has become an emerging issue of Asian elephant conservation.

海南长臂猿

拍摄地：海南省
摄影：武明录
Hainan Gibbon
Nomascus hainanus
Location: Hainan Province
Photographer: Wu Minglu

国家一级保护野生动物

海南长臂猿母子。

2021 年 3 月，海南长臂猿 2 个种群各新增一只婴猿。持续的监测显示这两只婴猿健康状况良好，生长发育正常。至此，海南热带雨林国家公园内的长臂猿野外种群数量从 40 年前的仅存 2 群不足 10 只增长到 5 群 35 只。

State First-class Protected Wild Animal

Hainan gibbon and its male cub.

In March 2021, a baby gibbon was added to each of the two Hainan gibbon populations. Continuous monitoring showed that the two cubs were in good health and normal growth. As a result, the wild population of Hainan gibbons in Hainan Tropical Rainforest National Park has expanded from 2 tribes with less than 10 individuals 40 years ago to 5 tribes with 35 ones.

海南长臂猿

拍摄地：海南省
摄影：张芬耀
Hainan Gibbon
Nomascus hainanus
Location: Hainan Province
Photographer: Zhang Fenyao

国家一级保护野生动物

在树上休息的雄性海南长臂猿。

海南长臂猿仅分布在中国海南省，是全世界最濒危的灵长类动物。

State First-class Protected Wild Animal

A male Hainan gibbon rests on the tree.

Hainan gibbon endemic to Hainan Province, and is one of the world's most endangered primates.

高黎贡白眉长臂猿

拍摄地：云南省
摄影：谢建国
Gaoligong Hoolock Gibbon
Hoolock tianxing
Location: Yunnan Province
Photographer: Xie Jianguo

国家一级保护野生动物

云南铜壁关省级自然保护区，两只高黎贡白眉长臂猿在密林枝丫上行走攀飞。

高黎贡山白眉长臂猿是唯一由中国科学家命名的类人猿。该物种主要分布于在高黎贡山区域，种群数量不超过 200 只。

State First-class Protected Wild Animal

Two Gaoligong hoolock gibbons swing from branches to branches in the dense forest of Tongbiguan Provincial Nature Reserve, Yunnan Province.

The Gaoligong hoolocks gibbon is the only ape named by the Chinese scientist. This species inhabits the Gaoligong Mountain area with a population of less than 200 individuals.

东黑冠长臂猿

拍摄地：广西壮族自治区

摄影：韦绍干

Cao-vit Crested Gibbon

Nomascus nasutus

Location: Guangxi Zhuang Autonomous Region

Photographer: Wei Shaogan

国家一级保护野生动物

东黑冠长臂猿雌雄异色，雄性黑色，雌性黄色。在中国，东黑冠长臂猿分布于中国广西与越南交界的一片喀斯特山林中，目前全球仅存 110 只左右，中国境内约有 33 只。

State First-class Protected Wild Animal

Cao-vit crested gibbons display sexual dimorphism, with the male in black and the female in yellow. In China, the cao-vit crested gibbons inhabit a karst forest at the border between Guangxi Zhuang Autonomous Region and Vietnam. At present, there are only about 110 cao-vit crested gibbons remaining in the world and about 33 ones in China.

北白颊长臂猿

拍摄地：云南省

摄影：陈建伟

Northern White-cheeked Gibbon

Nomascus leucogenys

Location: Yunnan Province

Photographer: Chen Jianwei

国家一级保护野生动物

被救助的北白颊长臂猿在人们的呵护下自在的生活。

北白颊长臂猿栖息于中、老、越三国交界地区的热带雨林和亚热带季雨林。在中国仅分布于云南，野外已十分罕见。

State First-class Protected Wild Animal

The rescued northern white-cheeked gibbons live freely under the proper care.

The northern white-cheeked gibbons inhabit the tropical rain forest and subtropical seasonal rain forest in the border area of China, Laos and Vietnam. The wild population of northern white-cheeked gibbon is so rare that, in China, it could only be found in Yunnan Province.

印支灰叶猴

拍摄地：云南省
摄影：武明录
Indochinese Gray Langur
Trachypithecus crepusculus
Location: Yunnan Province
Photographer: Wu Minglu

国家一级保护野生动物

印支灰叶猴在无量山和哀牢山两个国家级自然保护区内均有分布。近年来，当地不断加大生态建设和保护力度，印支灰叶猴的生存环境大为改善。

State First-class Protected Wild Animal

Indochinese gray langurs are distributed in two national nature reserves of Wuliang Mountain and Ailao Mountain. In recent years, with continuing strengthened local ecological construction and protection efforts, the habitat of the Indochinese gray langur has been greatly improved.

白头叶猴

拍摄地：广西壮族自治区
摄影：蒙有蔚
White-headed Langur
Trachypithecus leucocephalus
Location: Guangxi Zhuang Autonomous Region
Photographer: Meng Youwei

国家一级保护野生动物

崇左，白头叶猴集群栖息于中国广西南部的喀斯特地区，是由中国人命名的灵长类动物。经过持续努力，其野外种群已从20世纪80年代的300多只恢复到1300多只。

State First-class Protected Wild Animal

The white-headed langur inhabits in Chongzuo, the karst area in southern Guangxi Zhuang Autonomous Region, and is the primate species named by the Chinese researcher. After long-term conservation efforts, the number of wild populations has recovered from over 300 in 1980s to over 1,300.

黑叶猴

拍摄地：贵州省
摄影：谢建国
Francois's Langur
Trachypithecus francoisi
Location: Guizhou Province
Photographer: Xie Jianguo

国家一级保护野生动物

雌性黑叶猴携子穿越石桥，这里的黑叶猴每天在两岸林间觅食栖息。贵州麻阳河黑叶猴国家级自然保护区有大约 72 群 500 余只黑叶猴，是黑叶猴种群分布最密集，数量最多的地区。

State First-class Protected Wild Animal

A female Francois's Langur carries its cub across the stone bridge in Mayanghe, Guizhou Province. The Francois's Langur here forage and inhabit the forests on both sides of the bank every day.

The Mayanghe Francois's Langur National Nature Reserve which has about 72 tribes of more than 500 Francois's Langur is the area with the densest population and the largest number of Francois's Langur.

菲氏叶猴

拍摄地：云南省

摄影：郑山河

Phayre’s Langur

Trachypithecus phayrei

Location: Yunnan Province

Photographer: Zheng Shanhe

国家一级保护野生动物

菲氏叶猴长相萌动滑稽，毛色银灰，白眼圈、白嘴环，幼仔毛发为金色，十分可爱。

云南德宏芒市曾发现 5 群 320 只菲氏叶猴，是目前国内在单一地区发现的最大菲氏叶猴种群。

State First-class Protected Wild Animal

The Phayre’s langur looks cute and funny, with silver-grey fur, white eyes and white mouth. The cub has golden hair and looks adorable.

Five tribes of 320 Phayre’s langurs were found in Mangshi city, Dehong Prefecture, Yunnan Province, which is currently the largest population of Phayre’s langurs found in a single area in China.

川金丝猴

拍摄地：湖北省

摄影：徐永春

Golden Snub-nosed Monkey

Rhinopithecus roxellana

Location: Hubei Province

Photographer: Xu Yongchun

国家一级保护野生动物

湖北省神农架国家公园试点区，两只川金丝猴在树林中嬉戏。

神农架是川金丝猴分布最东端的地方，目前栖息有1470余只川金丝猴。

State First-class Protected Wild Animal

Two golden snub-nosed monkeys play in a forest in Shennongjia National Park pilot area, Hubei Province.

Shennongjia is the eastern boundary of golden snub-nosed monkeys' distribution, with more than 1,470 individuals at present.

滇金丝猴

拍摄地：云南省
摄影：于凤琴
Yunnan Snub-nosed Monkey
Rhinopithecus bieti
Location: Yunnan Province
Photographer: Yu Fengqin

国家一级保护野生动物

滇金丝猴的视觉非常发达，辨别色彩的能力很强。

每到春暖花开之际，滇金丝猴很喜欢到海拔3000米以上的杜鹃林活动。这株团花杜鹃树，是众多滇金丝猴的最爱，只有群中地位最高的家庭，才可以在这株杜鹃树上栖息。

State First-class Protected Wild Animal

The Yunnan snub-nosed monkey has a well-developed eye vision with great capability in distinguishing colors.

Every spring, Yunnan snub-nosed monkeys like to go to the rhododendron forest above 3,000 meters above sea level. This rhododendron (*Rhododendron anthosphaerum*) tree is the favorite of many Yunnan snub-nosed monkeys, but only the highest-ranking family in the tribe can inhabit this tree.

滇金丝猴

拍摄地：云南省
摄影：冯江
Yunnan Snub-nosed Monkey
Rhinopithecus bieti
Location: Yunnan Province
Photographer: Feng Jiang

国家一级保护野生动物

雪后的白马雪山，滇金丝猴在覆裹着积雪的枝丫间觅食。

滇金丝猴栖息于海拔2500~5000米的高山，是世界上栖息海拔最高的灵长类，被称为“雪山精灵”。

State First-class Protected Wild Animal

Yunnan snub-nosed monkeys forage among snow-covered branches on Baima Snow Mountain.

The Yunnan snub-nosed monkeys inhabit in mountains at an altitude of 2,500 to 5,000 meters. They are the primates with the highest habitat in the world and so called “the Elf of Snow Mountain”.

猕猴

拍摄地：广东省

摄影：徐永春

Rhesus Macaque

Macaca mulatta

Location: Guangdong Province

Photographer: Xu Yongchun

国家二级保护野生动物

广东内伶仃岛猕猴保护区，清晨，一群猕猴在海边觅食。距离深圳 9 千米的内伶仃岛是深圳最大的岛屿，岛上生活着超过 1000 只猕猴。

State Second-class Protected Wildlife

In the Neilingding Island Macaque Nature Reserve, Guangdong Province, a group of rhesus macaques foraging by the sea in the early morning. Neilingding Island, 9 kilometers away from Shenzhen, is the largest island in Shenzhen and home to more than 1,000 rhesus macaques.

东北虎

拍摄地：黑龙江省

摄影：谢建国

Amur Tiger

Panthera tigris altaica

Location: Heilongjiang Province

Photographer: Xie Jianguo

国家一级保护野生动物

黑龙江海林，森林中的东北虎。金秋季节，一只东北虎迎面而来，彰显虎威雄风。

State First-class Protected Wild Animal

An Amur tiger in the forest in Hailin County, Heilongjiang Province. In autumn, an oncoming Amur tiger demonstrates its mighty power.

雪豹

拍摄地：青海省
摄影：鲍永清
Snow Leopard
Panthera uncia
Location: Qinghai Province
Photographer: Bao Yongqing

国家一级保护野生动物

青海祁连山，一只雪豹在山岩间活动，毛色与环境融为一体。

State First-class Protected Wild Animal

A snow leopard blends in with its surroundings as moving among the rocks in Qilian Mountain, Qinghai Province.

雪豹

拍摄地：青海省

摄影：杰德·威恩嘉顿

Snow Leopard

Panthera uncia

Location: Qinghai Province

Photographer: Jed Weingarten

国家一级保护野生动物

青藏高原，一只雪豹在山脊上寻找猎物。雪豹主要以岩羊、盘羊等高原动物为主要食物。

State First-class Protected Wild Animal

A snow leopard is looking for prey on a ridge on the Tibetan Plateau. Snow leopards feed mainly on plateau animals such as blue sheep and argali.

猞猁

拍摄地：青海省
摄影：闹布战斗
Lynx
Lynx lynx
Location: Qinghai Province
Photographer: Norbu Dgradul

国家二级保护野生动物

猞猁母子回头张望的瞬间实为难得。

State Second-class Protected Wild Animal

A rare moment for a female lynx and its cubs looking back together.

兔狲

拍摄地：青海省

摄影：斯塔凡 · 威斯特兰德

Pallas's Cat

Felis manul

Location: Qinghai Province

Photographer: Staffan Widstrand

国家二级保护野生动物

摄影师在海拔 5000 米的地方，用长镜头拍摄了一组兔狲的图片，使这个憨态可掬的野生动物一时间成为中国“网红”。

State Second-class Protected Wild Animal

The photographer took a set of pictures of Pallas's Cat with a long lens at an altitude of 5,000 meters, making this charming animal an "Internet celebrity" in China for a time.

荒漠猫

拍摄地：青海省
摄影：尕布藏才郎
Chinese Mountain Cat
Felis bieti
Location: Qinghai Province
Photographer: Gasang Tsarang

国家一级保护野生动物

荒漠猫是中国特有的小型猫科动物，十分罕见。近年来，随着生态环境逐年向好，荒漠猫也开始出现在人们的视野中。

State First-class Protected Wild Animal

The Chinese mountain cat is a rare small feline endemic to China. In recent years, with the improvement of ecological environment in China, Chinese mountain cats have begun to appear in people's vision.

豹猫

拍摄地：云南省

摄影：赵建英

Leopard Cat

Prionailurus bengalensis

Location: Yunnan Province

Photographer: Zhao Jianying

国家二级保护野生动物

豹猫在中国也被称作“钱猫”，因为其身上的斑点很像中国的铜钱。体形和家猫相仿，但更加纤细，腿更长。

State Second-class Protected Wild Animal

The leopard cat is also called “coin cat” in China due to the ancient copper coin-like spots on their bodies. The leopard cat’s size is about of a domestic cat, but slenderer with longer legs.

金猫

拍摄地：四川省
摄影：肖飞
Asian Golden Cat
Catopuma temminckii
Location: Sichuan Province
Photographer: Xiao Fei

国家一级保护野生动物

摄影师在四川唐家河国家级保护区偶遇金猫母子，用相机记录下了这难得的瞬间。

金猫是一种外观变化非常大的动物，其毛色变化之多样在亚洲的猫科动物里绝无仅有，图为花斑型金猫。

State First-class Protected Wild Animal

The photographer captured the rare moment when coming across an Asian golden cat mother and its baby at Tangjiahe National Nature Reserve in Sichuan Province.

The appearances of Asian golden cat are varied greatly. The diversity of coat color is unique among the Asian cat species. The picture shows the spotted Asian golden cat.

亚洲胡狼

拍摄地：西藏自治区

摄影：彭建生

Golden Jackal

Canis aureus

Location: Tibet Autonomous

国家二级保护野生动物

2018 年，在西藏喜马拉雅山脉中段区域，摄影师拍摄到一只亚洲胡狼。这是首次通过影像确认有亚洲胡狼在中国分布。

State Second-class Protected Wild

In 2018, a photographer took the picture of a golden jackal in the middle area of the Himalayas in Tibet Autonomous Region. This is the first golden jackal image confirming species distribution in

赤狐

拍摄地：青海省

摄影：鲍永清

Red Fox

Vulpes vulpes

Location: Qinghai Province

Photographer: Bao Yongqing

国家二级保护野生动物

青海天骏，哺乳中的赤狐母子。

State Second-class Protected Wild Animal

Red fox is nursing babies in Tianjun, Qinghai Province.

喜马拉雅旱獭与藏狐

拍摄地：青海省
摄影：鲍永清
Himalayan Marmot and Tibetan Fox
Marmota himalayana & *Vulpes ferrilata*
Location: Qinghai Province
Photographer: Bao Yongqing

青海祁连山国家公园试点区，藏狐和旱獭对峙。旱獭是藏狐的捕食对象。

Confrontation between a Tibetan fox and a Himalayan marmot in the Qinghai Qilian Mountain National Park pilot area. Marmots are preys to Tibetan foxes.

藏原羚与狼

拍摄地：青海省
摄影：同海元
Tibetan Gazelle and Wolf
Procapra picticaudata and
Canis lupus
Location: Qinghai Province
Photographer: Tong Haiyuan

国家二级保护野生动物

青海格尔木，一头狼正在追逐捕食藏原羚幼仔。

State Second-class Protected Wild Animal

A wolf is chasing a Tibetan gazelle cub in Golmud, Qinghai Province.

貉

拍摄地：上海市
摄影：王放
Raccoon Dog
Nyctereutes procyonoides
Location: Shanghai
Photographer: Wang Fang

国家二级保护野生动物

夏夜，两只貉从上海一栋小区居民楼底部的通风口探出身来，谨慎地观察着周围的环境，准备开始自己的“夜生活”。近几年来，上海市野生貉数量迅速增长，种群数量推测有 5000 余只，成为上海城市野生动物管理不得不面对的一个问题。

State Second-class Protected Wild Animal

Two raccoon dogs emerged from a vent at the bottom of a residential building in Shanghai in summer night, cautiously watching surroundings as when preparing a “night life”. In recent years, the number of wild raccoon dogs has increased rapidly in Shanghai, with an estimated population of more than 5,000 individuals, which has become a problem for urban wildlife management in Shanghai.

水獭

拍摄地：青海省
摄影：闹布战斗
Eurasian Otter
Lutra lutra
Location: Qinghai Province
Photographer: Norbu Dgradul

国家二级保护野生动物

水獭主要栖息于河流和湖泊一带，尤其喜欢生活在两岸林木繁茂的溪河地带，分布范围极广，亚洲、欧洲、非洲都有其踪迹。

State Second-class Protected Wild Animal

Eurasian otters mainly inhabit rivers and lakes, especially prefer to live in rivers with lush trees on both sides. They are widely distributed in Asia, Europe and Africa.

藏羚

拍摄地：青海省
摄影：果洛 · 索南
Tibetan Antelope
Pantholops hodgsonii
Location: Qinghai Province
Photographer: Golo Sonam

国家一级保护野生动物

青海可可西里，两只雄性藏羚在晨光中追逐争夺交配权。每年 12 月初至次年 1 月初是藏羚的交配期。由于国际市场对藏羚羊绒披肩的需求，藏羚一度被大量捕杀。经过 30 余年的保护，种群数量从 7 万只增加到 30 万只，藏羚受危程度从濒危降级为近危。

State First-class Protected Wild Animal

Two male Tibetan antelopes are fighting for the copulation in the morning light reflection in Hoh Xil, Qinghai Province. The mating season of Tibetan antelopes is from early December to early January next year. Due to the international market demand for the shahtoosh, Tibetan antelopes were once seriously poached. After more than 30 years of conservation in China, the population has increased from 70,000 to 300,000, and the status of Tibetan antelope has been downgraded from “endangered” to “near threatened” of *IUCN Red List*.

藏羚

拍摄地：青海省
摄影：奚志农
Tibetan Antelope
Pantholops hodgsonii
Location: Qinghai Province
Photographer: Xi Zhinong

国家一级保护野生动物
可可西里国家级自然保护区，一轮明月下的藏羚羊。
这里是中国、面积最大、海拔最高，野生动物资源最丰富的自然保护区之一。

State First-class Protected Wild Animal
Tibetan antelopes are under the bright moon in Hoh Xil National Nature Reserve, which is one of the nature reserves with the largest area, the highest altitude and the richest wildlife resources in China.

普氏原羚

拍摄地：青海省

摄影：徐永春

Przewalski's Gazelle

Procapra przewalskii

Location: Qinghai Province

Photographer: Xu Yongchun

春节求偶季节，争斗的雄性普氏原羚。

State First-class Protected Wild Animal

Male Przewalski's gazelles are fighting for copulation during mating season in spring.

盘羊

拍摄地：青海省
摄影：斯塔凡·威斯特兰德
Argali
Ovis ammon
Location: Qinghai Province
Photographer: Staffan Widstrand

国家一级保护野生动物

盘羊是盘羊属中体形最大的山地动物，在中国主要分布在西部地区。雄性的弯角粗大，最长可达 1 米以上，向下扭曲呈螺旋状，外侧有环棱，可判断年龄；雌性的角则较短，而且弯度不大。

State Second-class Protected Wild Animal

The argali is the largest mountain animal in the genus of *Ovis*, mainly distributing in the western region of China. Males have thick curved horns, up to one meter long. The horn is downward twist spiral shape, with outer ring edges, which can be used to determine age. Horns of females are shorter and less curved.

塔尔羊

拍摄地：西藏自治区
摄影：彭建生
Himalayan Tahr
Hemitragus jemlahicus
Location: Tibetan Autonomous Region
Photographer: Peng Jiansheng

国家一级保护野生动物

秋日的珠峰西侧峡谷，成年雄性塔尔羊进入求偶期，毛色光亮威武，雄姿勃发。

State First-class Protected Wild Animal

On the west side of Mountains Qomolangma Canyon in autumn, the adult male Himalayan tahr starts its courtship period, with bright fur and vigorous posture.

岩羊

拍摄地：四川省
摄影：彭建生
Blue Sheep
Pseudois nayaur
Location: Sichuan Province
Photographer: Peng Jiansheng

国家二级保护野生动物

秋意正浓，一群在山林间觅食的岩羊。岩羊主要栖息于2100~6300米陡峭的高山裸岩地带，以善于攀岩而得名，是雪豹的主要食物。

State Second-class Protected Wild Animal

In autumn, a flock of blue sheep forage in the forests and mountains. The blue sheep mainly inhabits the steep bare rock zone of 2,100 to 6,300 meters. They are named “rock sheep” in Chinese for their good rock-climbing ability. The blue sheep is the main prey species of snow leopards.

秦岭羚牛

拍摄地：陕西省
摄影：孙晋强
Golden Takin
Budorcas bedfordi
Location: Shaanxi Province
Photographer: Sun Jinqiang

国家一级保护野生动物

陕西牛背梁国家级自然保护区，在云雾缭绕的秦岭主脊觅食休息的秦岭羚牛。

State First-class Protected Wild Animal

In Shaanxi Niubeiliang National Nature Reserve, golden takins forage and rest on the mist-shrouded main ridge of the Qinling mountains.

贡山羚牛

拍摄地：云南省
摄影：彭建生
Gongshan Takin
Budorcas taxicolor
Location: Yunnan Province
Photographer: Peng Jiansheng

国家一级保护野生动物
在溪流旁休息的高黎贡山羚牛。
相比秦岭羚牛，高黎贡山羚牛为棕褐色和淡金黄色，颜色较深。
State First-class Protected Wild Animal
The Gongshan takin rest by the stream.
Compared with the golden takin, the fur of Gongshan takin is daker with brown and light golden color.

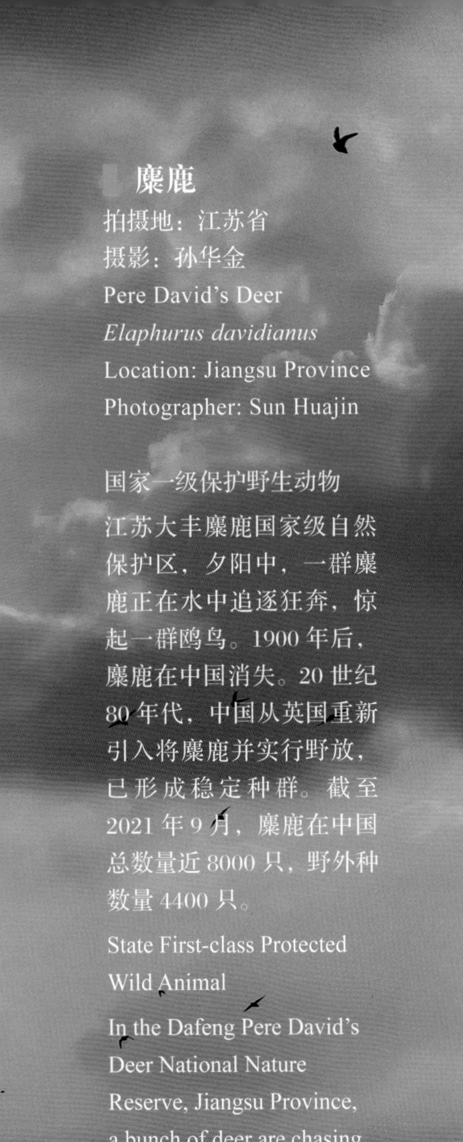

麋鹿

拍摄地：江苏省

摄影：孙华金

Pere David's Deer

Elaphurus davidianus

Location: Jiangsu Province

Photographer: Sun Huajin

国家一级保护野生动物

江苏大丰麋鹿国家级自然保护区，夕阳中，一群麋鹿正在水中追逐狂奔，惊起一群鸥鸟。1900 年后，麋鹿在中国消失。20 世纪 80 年代，中国从英国重新引入将麋鹿并实行野放，已形成稳定种群。截至 2021 年 9 月，麋鹿在中国总数量近 8000 只，野外种数量 4400 只。

State First-class Protected Wild Animal

In the Dafeng Pere David's Deer National Nature Reserve, Jiangsu Province, a bunch of deer are chasing and running in the water in the sunset, startled a flock of gulls. Since 1900, the Pere David's deer locally extinct in China. In 1980s, a re-introduced population from the United Kingdom, and formed a stable population. By September 2021, the total number of Pere David's deer in China is nearly 8,000, with a wild population of 4,400.

坡鹿

拍摄地：海南省
摄影：顾晓军
Eld's Deer
Rucervus eldii
Location: Hainan Province
Photographer: Gu Xiaojun

国家一级保护野生动物

坡鹿因喜欢栖息在丘陵草坡而得名。曾广布于海南岛全境，但到20世纪70年代中期，海南坡鹿只剩下约40头。经过几十年的严格保护，数量已经达到1000多头，一个行将灭绝的物种又重新恢复了生机，主要栖息在海南邦溪省级坡鹿保护区。

State First-class Protected Wild Animal

The Eld's deer is also named the "slope deer" in Chinese because it prefers to live on the hilly grass slopes. They were once widely distributed throughout Hainan Island, but by the mid-1970s, there were only about 40 individuals left. After decades of rigorous conservation, the number of Eld's deer has reached more than 1,000, and the species that was once about to be extinct has been revived. It mainly inhabits the Bangxi Provincial Eld's Deer Reserve in Hainan Province.

梅花鹿

拍摄地：山东省
摄影：宋林继
Sika Deer
Cervus nippon
Location: Shandong Province
Photographer: Song Linji

国家一级保护野生动物

海岛精灵——野生梅花鹿

生活在山东威海刘公岛上的梅花鹿，是 20 世纪 70 年代初期从四川引进的。经过多年的保护野化，刘公岛的野生梅花鹿已从当初 7 只，增加到如今的 200 只左右。

State First-class Protected Wild Animal

Island elf—Wild Sika deer

Sika deer, which live on Liugong Island in Weihai, Shandong Province, were introduced from Sichuan Province in the early 1970s. After years of rewilding and conservation, the number of wild sika deer in Liugong Island has increased from 7 to about 200 today.

白唇鹿

拍摄地：青海省

摄影：宋林继

White-lipped Deer

Przewalskium albirostris

Location: Qinghai Province

Photographer: Song Linji

国家一级保护野生动物

2020 年 12 月 16 日，青海省曲麻莱县通天河流域，大群的白唇鹿从海拔 4200 米的高山上奔跑而下。

State First-class Protected Wild Animal

On 16 December 2020, a large herd of white-lipped deer ran down from a high mountain of 4,200 meters elevation in Tongtian River Basin, Qumalai County, Qinghai Province.

马鹿

拍摄地：内蒙古自治区

摄影：武明录

Red Deer

Cervus elaphus

Location: Inner Mongolia Autonomous Region

Photographer: Wu Minglu

国家二级保护野生动物

阿尔山白桦林间觅食的雄性马鹿。

State Second-class Protected Wild Animal

A male red deer forages in the birch forest in Arshan Mountain.

野骆驼
拍摄地：甘肃省
摄影：徐永春
Bactrian Camels
Camelus ferus
Location: Gansu Province
Photographer: Xu Yongchun

国家一级保护野生动物

罗布泊野骆驼国家级自然保护区，发情期的野骆驼。

野骆驼数量极其稀少，世界上分布有野骆驼的国家只有中国和蒙古，种群数量不足1000头，中国分布有650头左右。

State First-class Protected Wild Animal

A bactrian camel in estrus in Lop Nur Wild Camel National Nature Reserve.

The bactrian camel is extremely endangered. The countries in the world with wild bactrian camel population are China and Mongolia, with total number of less than 1,000, among those about 650 in China.

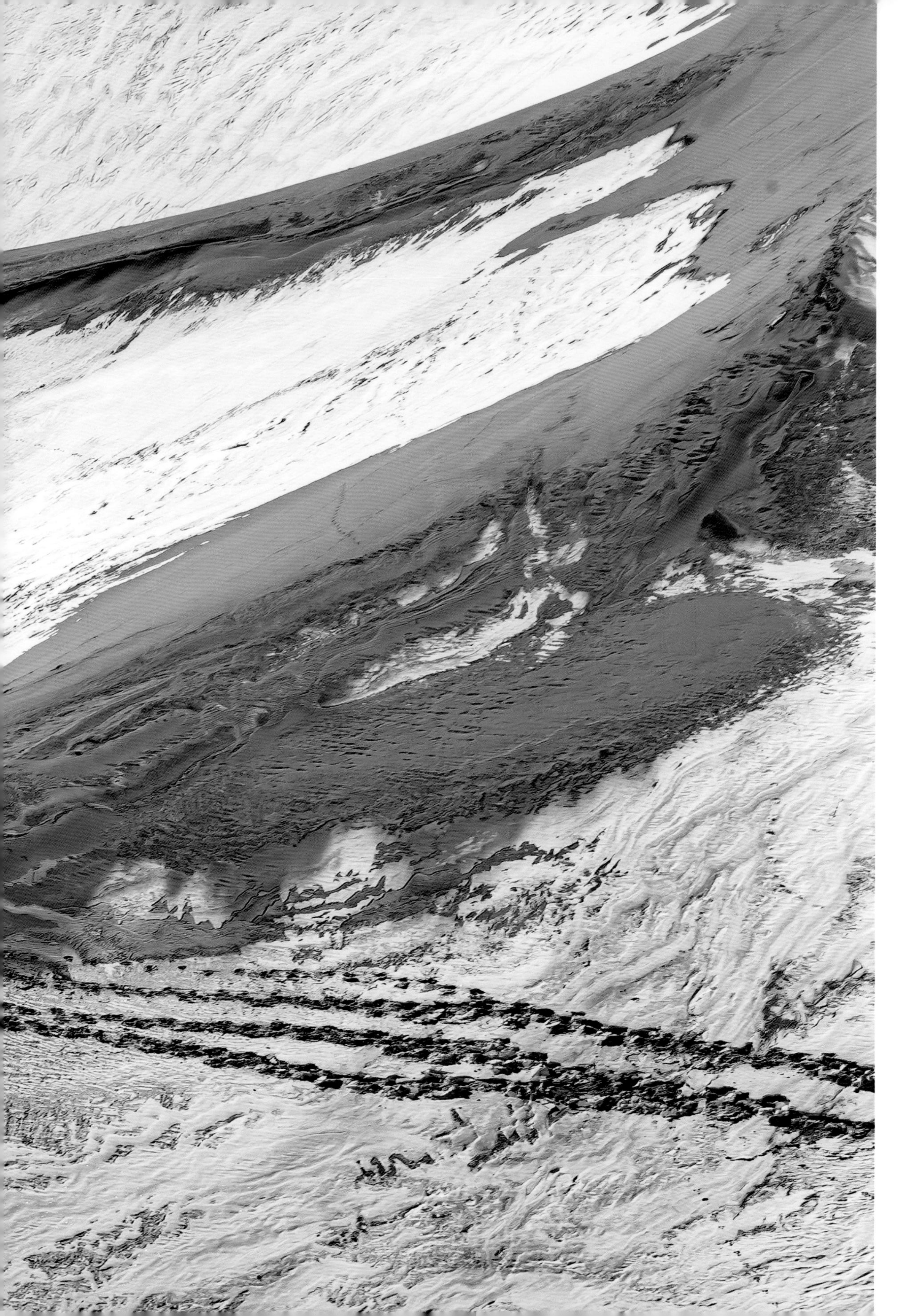

野牦牛

拍摄地：青海省
摄影：何启金
Wild Yak
Bos mutus
Location: Qinghai Province
Photographer: He Qijin

国家一级保护野生动物

栖息于青海可可西里的野牦牛。成群的野牦牛会主动逃避敌害，而性情凶狠暴戾的孤牛常会主动攻击在它面前经过的各种对象。

State First-class Protected Wild Animal

A wild yak in Hoh Xil, Qinghai Province. Herds of wild yaks would take the initiative to avoid the enemy. Fierce and violent solitary yak often actively attack any objects passing in front.

普氏野马

拍摄地：甘肃省
摄影：徐永春
Przewalski's Horse
Equus ferus
Location: Gansu Province
Photographer: Xu Yongchun

国家一级保护野生动物

普氏野马是我国自然分布的物种，曾一度在野外消失，经过多年不懈努力，人工繁育种群不断壮大，并重建了野外种群。我国现存普氏野马种群数量已突破 700 匹。

State First-class Protected Wild Animal

China is the natural range of Przewalski's horses. The species once disappeared in the wild. After years of unremitting efforts, the captive breeding population has been growing, and the wild population has been rebuilt with more than 700 individuals in China.

藏野驴

拍摄地：新疆维吾尔自治区
摄影：雷洪
Kiang
Equus kiang
Location: Xinjiang Uygur Autonomous Region
Photographer: Lei Hong

国家一级保护野生动物

阿尔金山国家级自然保护区，一群藏野驴在草地上飞奔。阿尔金山深处的草原、雪山、沙漠、湖泊等是众多野生动物的乐园。

State First-class Protected Wild Animal

A herd of kiangs are running on the grass in the Altun Mountain National Nature Reserve. The grasslands, snow mountains, deserts and lakes in the depths of the Altun Mountains are the paradise to many wild animals.

棕熊

拍摄地：青海省
摄影：宋林继
Brown Bear
Ursus arctos
Location: Qinghai Province
Photographer: Song Linji

国家二级保护野生动物

青海格尔木国家级自然保护区，棕熊夜间带着两个幼仔外出觅食。棕熊食性较杂，有时会挖掘洞穴捕食鼠兔和旱獭，甚至闯入社区翻食居民家中食物，对人类构成威胁。

State Second-class Protected Wild Animal

A brown bear took two cubs out for foraging at night in Geermu National Nature Reserve, Qinghai. The diet of brown bears is quite diverse that sometimes they dig caves to prey on pikas and marmots, and even break into local communities for food from residents' houses, causing a threat to local people.

黑熊

拍摄地：四川省
摄影：邓建新
Asian Black Bear
Ursus thibetanus
Location: Sichuan Province
Photographer: Deng Jianxin

国家二级保护野生动物

四川唐家河国家级保护区，两只黑熊幼仔爬到树上吃野果。

State Second-class Protected Wild Animal

Two Asian black bear cubs climb onto a tree for fruits in Tangjiahe National Nature Reserve, Sichuan Province.

小熊猫

拍摄地：四川省

摄影：斯塔凡・威斯特兰德

Red Panda

Ailurus fulgens

Location: Sichuan Province

Photographer: Staffan Widstrand

国家二级保护野生动物

小熊猫与大熊猫并非近亲，却有着相似的栖息地和食性。由于栖息地退化等因素，小熊猫的受威胁等级在中国被评估为易危（VU），为保护大熊猫而建立的许多自然保护区也为该物种提供了庇护。

State Second-class Protected Wild Animal

Red pandas are not phylogenetically related to giant pandas, but they share similar habitats and diets. Due to threats such as habitat degradation, the conservation status of red pandas has been assessed as “vulnerable (VU)” in China, and many nature reserves established for giant panda conservation also benefit this species.

伊犁鼠兔

拍摄地：新疆维吾尔自治区
摄影：李维东
Ili Pika
Ochotona iliensis
Location: Xinjiang Uygur Autonomous Region
Photographer: Li Weidong

国家二级保护野生动物

四处张望的伊犁鼠兔。伊犁鼠兔是1986年发现命名的兔形目新种动物，数量极为稀少。这张照片中，萌动可爱的伊犁鼠兔惊艳了世界，迅速引起各界的关注。2021年，伊犁鼠兔被列为国家二级保护野生动物。

State Second-class Protected Wild Animal

An Ili pika is looking around. The Ili pika is a new species of Lagomorpha, discovered and named in 1986 with extremely small population. In this photo, the adorable Ili pika has surprised the world and quickly attracted attention from all aspects. In 2021, it was listed as State Second-class Protected Wild Animal.

巨松鼠

拍摄地：云南省
摄影：赵建英
Black Giant Squirrel
Ratufa bicolor
Location: Yunnan Province
Photographer: Zhao Jianying

国家二级保护野生动物

巨松鼠是一种大型啮齿类动物，栖息于海拔 2000 米以下的热带、亚热带季雨林的高树上，是典型的树栖动物。

State Second-class Protected Wildlife

The black giant squirrel is a kind of large rodent and a typical arboreal animal, which inhabits high trees in tropical and subtropical seasonal rain forests below 2,000 meters elevation.

中华穿山甲

拍摄地：浙江省
摄影：周佳俊
Chinese Pangolin
Manis pentadactyla
Location: Zhejiang Province
Photographer: Zhou Jiajun

国家一级保护野生动物

中华穿山甲是特化物种，视觉基本退化、嗅觉灵敏，以蚁类为食。生境丧失和破碎化、非法贸易等对穿山甲的生存造成威胁。

State First-class Protected Wild Animal

The Chinese pangolin is a specialized species with degraded vision and sensitive sense of smell, and feeding on ants. Habitat loss and fragmentation, and illegal trade threaten the survival of Chinese pangolins.

▍白尾海雕

拍摄地：河北省

摄影：徐永春

White-tailed Sea Eagle

Haliaeetus albicilla

Location: Hebei Province

Photographer: Xu Yongchun

国家一级保护野生动物

2011 年北戴河，被当地救助的一只白尾海雕，放飞前的瞬间，仍显其空中霸主神威。

State First-class Protected Wild Animal

In 2011, a rescued white-tailed sea-eagle showed its prestige at the moment right before being released in Beidaihe, Hebei Province.

鸟类

BIRDS

中国是世界上鸟类资源比较丰富的国家之一，现有鸟类 1445 种，约占世界鸟类总种数的 14%。物种多样性高、特有种丰富、区系起源古老，是中国鸟类资源的三大特点。由于有将近一半的鸟类属于长距离迁徙的候鸟，因此中国鸟类资源的保护不仅关系到中国自身的生态安全，而且对保护全球珍稀濒危鸟类、维持全球的生物多样性都具有重要的意义。

China is rich in bird species, with 1,445 species of birds (14% of global bird species). Many birds in China are of ancient origin, as well as endemic to China. As more than a half of the bird species are long-distance migratory, bird conservation is not only concerning China, but also contributes to the global biodiversity conservation.

黑颈鹤

拍摄地：西藏自治区
摄影：栾亚立
Black-necked Crane
Grus nigricollis
Location: Tibet Autonomous Region
Photographer: Luan Yali

国家一级保护野生动物

春暖花开时节，一群黑颈鹤在西藏林芝迁徙地湖边驻留，桃花湖水相衬映，如同仙境。

State First-class Protected Wild Animal

In warm spring, a flock of black-necked cranes stay by the lake in the migratory area in Nyingchi, Tibet Autonomy Region. The black-necked cranes contrast with peach blossoms and lake water, forming an image like a fairyland.

丹顶鹤

拍摄地：黑龙江省

摄影：武明录

Red-crowned Crane

Grus japonensis

Location: Heilongjiang Province

Photographer: Wu Minglu

国家一级保护野生动物

黑龙江扎龙国家级自然保护区，朝霞中飞来觅食的丹顶鹤，如大自然灵动的舞者。

State First-class Protected Wild Animal

In Zhalong National Nature Reserve, Heilongjiang Province, red-crowned cranes flying to forage in the morning glow are just like graceful dancers of nature.

黑颈鹤

拍摄地 ：西藏自治区
摄影 ：栾亚立
Black-necked Crane
Grus nigricollis
Location: Tibet Autonomous Region
Photographer: Luan Yali

国家一级保护野生动物

飞越喜马拉雅山脉的黑颈鹤。

黑颈鹤是唯一主要生活在高原并主要分布在中国的鹤。栖息地的保护正在帮助黑颈鹤种群稳定增长，种群数量达 10000 多只。

State First-class Protected Wild Animal

Black-necked cranes flying over the Himalayas.

Black-necked cranes are the only cranes that mainly live on plateaus in China. The habitat protection has contributed to helping the steady growth of black-necked crane population, which exceeds 10,000 individuals.

丹顶鹤

拍摄地：黑龙江省
摄影：赵建英
Red-crowned Crane
Grus japonensis
Location: Heilongjiang Province
Photographer: Zhao Jianying

国家一级保护野生动物

引颈高歌的丹顶鹤。

丹顶鹤在中国传统文化中有极高的知名度，是吉祥长寿的象征。中国建立了多个以保护丹顶鹤为主的自然保护区，有效地保护了丹顶鹤及其生存环境。

State First-class Protected Wild Animal

The singing red-crowned crane.

The red-crowned crane is very well-known in Chinese traditional culture as a symbol of auspiciousness and longevity. A number of nature reserves have been established in China for specially effective conservation of the red-crowned crane and its habitat.

▌蓑羽鹤

拍摄地：内蒙古自治区

摄影：王秀荣

Demoiselle Crane

Grus virgo

Location: Inner Mongolia Autonomous Region

Photographer: Wang Xiurong

国家二级保护野生动物

内蒙古克什克腾草原，正在觅食的蓑羽鹤一家与黄牛相遇，蓑羽鹤亲鸟张开翅膀奋力保护幼鸟。

State Second-class Protected Wild Animal

In the Keshiketeng Grassland of Inner Mongolia, a family of feeding demoiselle cranes met a cattle. The demoiselle crane parents spread their wings to protect their young.

白鹤

拍摄地：江西省

摄影：张春晓

Siberian Crane

Grus leucogeranus

Location: Jiangxi Province

Photographer: Zhang Chunxiao

国家一级保护野生动物

白鹤约有 4000 只在中国越冬，绝大多数选择在江西鄱阳湖，这是大自然给美丽中国的馈赠。

State First-class Protected Wild Animal

About 4,000 Siberian cranes overwinter in China. Most of them choose to stay in Poyang Lake in Jiangxi Province. Their presence is a precious gift from nature to beautiful China.

灰鹤

拍摄地：北京市
摄影：赵建英
Eurasian Crane
Grus grus
Location: Beijing
Photographer: Zhao Jianying

国家二级保护野生动物

北京野鸭湖湿地，飞翔的灰鹤把这里的冬季幻化成一个美丽的童话世界。

State Second-class Protected Wild Animal

The flying Eurasian cranes turn the winter of Beijing Yeyahu wetland into a beautiful fairyland.

朱鹮

拍摄地：陕西省
摄影：袁允
Crested Ibis
Nipponia nippon
Location: Shaanxi Province
Photographer: Yuan Yun

国家一级保护野生动物

在春天繁殖季节，一对朱鹮在晨光中高声鸣唱。

State First-class Protected Wild Animal

A pair of crested ibises are singing in the morning light during the spring breeding season.

疣鼻天鹅

拍摄地：新疆维吾尔自治区
摄影：谢建国

Mute swan
Cygnus olor
Location: Xinjiang Uygur Autonomous Region
Photographer: Xie Jianguo

国家二级保护野生动物

伊犁河北岸的新疆伊宁县英塔木乡天鹅泉湿地，因为水下有多处温泉，冬季水面不结冰，成为疣鼻天鹅的栖息越冬地。这里童话般的雾凇天鹅景观，令无数游人倾倒。

State Second-class Protected Wild Animal

The Swan Spring Wetland in Yingtamu Township, Yining County, Xinjiang Province, located on the north bank of the Yili River has many hot springs underwater, making the water surface unfrozen in winter. It becomes an overwintering habitat for mute swans. The fairy-tale rime and swan landscape here has attracted countless tourists.

小天鹅

拍摄地：内蒙古自治区
摄影：谢建国
Tundra Swan
Cygnus columbianus
Location: Inner Mongolia Autonomous Region
Photographer: Xie Jianguo

国家二级保护野生动物

内蒙古阿拉善腾格里沙漠，点缀着百余个大小不一、宁静清幽的湖泊。每年春秋候鸟迁徙季节，大、小天鹅，雁鸭等候鸟在此停留，形成大漠天鹅湖的壮美画面。

State Second-class Protected Wild Animal

The Alxa Tengger Desert in Inner Mongolia is dotted with more than a hundred peaceful lakes of various sizes. During the spring and autumn migratory season, whooper swans, tundra swans, geese and ducks stay here, forming a magnificent figure of the desert swan lake.

大天鹅

拍摄地：北京市
摄影：谢建国
Whooper Swan
Cygnus cygnus
Location: Beijing
Photographer: Xie Jianguo

国家二级保护野生动物

北京怀柔水库，雪后初晴，湖面冰雪融映成一朵朵花样图案，大天鹅水中游弋，好似天使落菊台。

State Second-class Protected Wild Animal

In Huairou Reservoir in Beijing, snow and ice melting into patterns of flowers, whooper swans swim in the water, like angels falling chrysanthemum terrace.

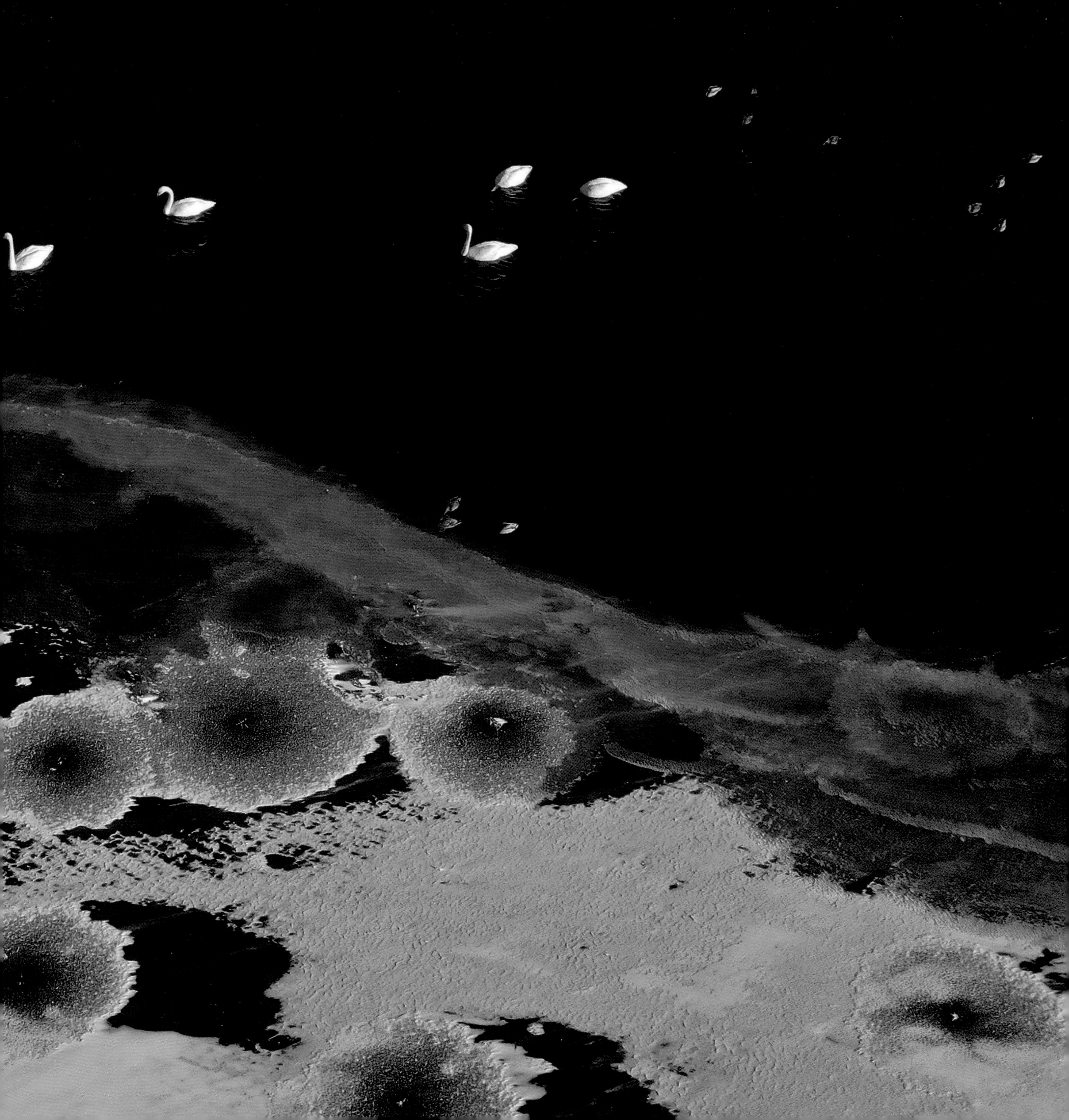

▌疣鼻天鹅
拍摄地：新疆维吾尔自治区
摄影：谢建国
Mute Swan
Cygnus olor
Location: Xinjiang Uygur Autonomous Region
Photographer: Xie Jianguo

国家二级保护野生动物

伊犁河谷天鹅泉，雪雾中，疣鼻天鹅羽翼高耸，玉珠垂挂，身姿素雅，冰清玉洁。

State Second-class Protected Wild Animal

Elegant posture of the mute swan in the snow and fog in the Swan Spring of the Yili River valley.

▌勺嘴鹬
拍摄地：海南省
摄影：熊跃辉
Spoon-billed Sandpiper
Calidris pygmeus
Location: Hainan Province
Photographer: Xiong Yuehui

国家一级保护野生动物

2020 年底摄影师在海南儋州新盈红树林湿地发现 6 只勺嘴鹬。勺嘴鹬在海南十分少见。

勺嘴鹬是世界上最濒危的鸟类之一，全球种群数量不足 600 只，江苏东台条子泥滩涂是它们迁徙路上最重要的停歇地之一。

State First-class Protected Wild Animal

At the end of 2020, photographer spotted six spoon-billed sandpipers in Xinying Mangrove Wetland in Danzhou, Hainan Province. Spoon-billed sandpipers are rare in Hainan.

The spoon-billed sandpiper is one of the most endangered birds in the world, with a global population of less than 600. The Tiaozini Marshland in Dongtai, Jiangsu Province is one of the most important stopover sites on their migration route.

鸳鸯

拍摄地：北京市

摄影：谢建国

Mandarin Duck

Aix galericulata

Location: Beijing

Photographer: Xie Jianguo

国家二级保护野生动物

初冬，玉渊潭湖面结了薄冰，夕阳辉映下，一群雄鸳鸯飞上冰面，在雌鸳鸯前炫耀争雄。

玉渊潭公园是北京城区水域面积最大的湿地，每年有100余种鸟儿在此繁衍生息。

State Second-class Protected Wild Animal

In early winter, a flock of male Mandarin ducks land on the thin ice formed on top of the Yuyuantan Lake in the sunset, courting to the females.

The Yuyuantan Park is the largest wetland in the urban area of Beijing, where more than 100 species of birds live and breed every year.

中华秋沙鸭

拍摄地：吉林省
摄影：王莅翔
Scaly-sided Merganser
Mergus squamatus
Location: Jilin Province
Photographer: Wang Lixiang

国家一级保护野生动物

中华秋沙鸭是第三纪冰川末期遗留下来的古老物种，素有鸟中“大熊猫”之称。

中国小兴安岭是中华秋沙鸭的重要繁殖地。每年春季，中华秋沙鸭在树洞中产卵孵化，雏鸟出壳后随亲鸟跳出树洞，在附近平缓河段活动。

State First-class Protected Wild Animal

The scaly-sided merganser is an ancient species left over from the end of the Tertiary Glacial Period, and is known as the “giant panda” among birds .

Xiao Hinggan Mountains is an important breeding site for scaly-sided mergansers. Every spring, the scaly-sided mergansers lay eggs in the tree holes, and the newly hatched chicks would follow the parent birds to jump out of the tree holes and move around in the gentle river nearby.

中华秋沙鸭

拍摄地：湖南省

摄影：赵建英

Scaly-sided Merganser

Mergus squamatus

Location: Hunan Province

Photographer: Zhao Jianying

国家一级保护野生动物

中华秋沙鸭主要在我国长江中下游越冬。湖南省沅江流域是中华秋沙鸭的重要越冬栖息地之一。

State First-class Protected Wild Animal

Scaly-sided mergansers winter mainly in the middle and lower reaches of Yangtze River in China. The Yuanjiang River Basin of Hunan Province is one of its important overwintering habitats.

河燕鸥

拍摄地：云南省
摄影：孙晓宏
River Tern
Sterna aurantia
Location: Yunnan Province
Photographer: Sun Xiaohong

国家一级保护野生动物

河燕鸥是中国数量最少的鸟类之一，仅见于云南大盈江流域。通过不断提升保护措施，为它们创造良好的生存环境，河燕鸥由 2019 年的 7 只增长到目前的 13 只。

State First-class Protected Wild Animal

The river tern is one of the least abundant birds in China and only found in the Dayingjiang River Basin in Yunnan Province. By continuously improving conservation measures and creating an ideal habitat, the number of river terns has grown from 7 in 2019 to 13 at present.

中华凤头燕鸥

拍摄地：福建省
摄影：陈林
Chinese Crested Tern
Thalasseus bernsteini
Location: Fujian Province
Photographer: Chen Lin

国家一级保护野生动物

中华凤头燕鸥，又名黑嘴端凤头燕鸥，被世界自然保护联盟列入极度濒危物种，是“最濒危的 100 个物种”之一，现今全球总数约 100 只。

State First-class Protected Wild Animal

Chinese crested tern, also known as black-billed crested tern, is listed as “critically endangered” by the International Union for Conservation of Nature (IUCN). It is also one of the “100 most endangered species”, with a global population of about 100.

遗鸥

拍摄地：天津市

摄影：徐永春

Relict Gull

Ichthyaetus relictus

Location: Tianjin

Photographer: Xu Yongchun

国家一级保护野生动物

天津海滨滩涂是遗鸥的重要越冬地，每年都有成千上万只遗鸥在这里越冬栖息。

State First-class Protected Wild Animal

Tianjin coastal beach is an important overwintering site for relict gulls. Thousands of relict gulls stay here every winter.

卷羽鹈鹕

拍摄地：宁夏回族自治区
摄影：张月琴
Dalmatian Pelican
Pelecanus crispus
Location: Ningxia Hui
Autonomous Region
Photographer: Zhang Yueqin

国家一级保护野生动物
在黄河里游弋的卷羽鹈鹕。
2020 年冬季，卷羽鹈鹕再度飞临银川黄河湿地，这是当地历史上第二次记录到卷羽鹈鹕的踪迹。
State First-class Protected Wild Animal
In the winter of 2020, the dalmatian pelicans flew to the Yellow River wetland in Yinchuan again, which was the second dalmatian pelican record in local history.

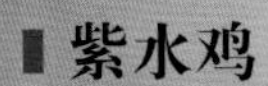

紫水鸡

拍摄地：云南省
摄影：徐永春
Purple Swamphen
Porphyrio porphyrio
Location: Yunnan Province
Photographer: Xu Yongchun

国家二级保护野生动物

春季身披艳丽繁殖羽的雄性紫水鸡为争夺配偶激战正酣。

State Second-class Protected Wild Animal

In spring, male purple swamphens with gorgeous breeding feathers are fighting for courtship.

黑脸琵鹭

拍摄地：辽宁省

摄影：武明录

Black-faced Spoonbill

Platalea minor

Location: Liaoning Province

Photographer: Wu Minglu

国家一级保护野生动物

黑脸琵鹭是东亚地区特有鸟类，2021 年，全球统计共有 5222 只，是世界上琵鹭属中分布范围最窄的物种。大连庄河形人坨岛是黑脸琵鹭全球重要的繁殖地。

State First-class Protected Wild Animal

The black-faced spoonbill is an endemic bird in East Asia. In 2021, 5,222 individuals were recorded worldwide. The distribution of this species is the narrowest among species in the genus *Platalea* around the world. Xingrentuo Island in Zhuanghe, Dalian is an important breeding site for black-faced spoonbills worldwide.

黑鹳

拍摄地：北京市

摄影：谢建国

Black Stork

Ciconia nigra

Location: Beijing

Photographer: Xie Jianguo

国家一级保护野生动物

捕获鲶鱼的黑鹳。

北京房山拒马河是黑鹳的主要栖息地，经过多方努力，目前这里的黑鹳种群数量由2000年的5只恢复到70余只。

State First-class Protected Wild Animal

A black stork catching catfish.

The Juma River in Fangshan, Beijing is the main habitat of black storks. Benefiting from significant efforts, the population of black storks here has recovered from 5 in 2000 to more than 70.

鸻鹬鸟浪

拍摄地：辽宁省

摄影：黄冬青

Waves of Shorebirds

Location: Liaoning Province

Photographer: Huang Dongqing

鸻鹬鸟浪形状如兔。

每年春季迁徙季，数十万只鸻鹬类水鸟大军从南半球飞抵鸭绿江口湿地自然保护区，再飞往西伯利亚和阿拉斯加等地。潮汐把鸟儿汇聚成群，万鸟齐飞，如变幻的沙画描绘出大自然神奇的美丽画卷。

A rabbit-like shape formed by waves of shorebirds.

Every spring, hundreds of thousands of shorebirds fly from the southern hemisphere to the Yalu River Estuary Wetland Nature Reserve, Liaoning Province, and then head to Siberia and Alaska. The tide gathers the birds in groups, and thousands of birds fly together, like a sand painting depicting a beautiful and magical beauty of nature.

▍黄腹角雉
拍摄地：广东省
摄影：冯江
Cabot's Tragopan
Tragopan caboti
Location: Guangdong Province
Photographer: Feng Jiang

国家一级保护野生动物

广东南岭国家级自然保护区，黄腹角雉雄鸟发情期喉下的肉裙膨胀，色彩更加炫目，头上一对黄色的肉角挺直锃亮。雄鸡面对雌鸟频频点头，挺身展翅又低头长叫，激情迸发。

中国是世界上雉鸡类种数最多的国家。

State First-class Protected Wild Animal

In Nanling National Nature Reserve, Guangdong Province, the fleshy skirt under the throat of the male Cabot's tragopan swells during estrus with more dazzling color. A pair of yellow fleshy horns on the head is straight and shiny. The male nods frequently, stretches his wings and crows with passion in the face of the female.

China has the highest number of pheasant species in the world.

▍黄腹角雉
拍摄地：江西省
摄影：赵建英
Cabot's Tragopan
Tragopan caboti
Location: Jiangxi Province
Photographer: Zhao Jianying

国家一级保护野生动物

江西三清山景区，雨后，黄腹角雉的幼鸟钻进母亲的翅膀下取暖。

State First-class Protected Wild Animal

A Cabot's tragopan chick flees under mother's wing to warm itself after rain in Sanqingshan Scenic Area, Jiangxi Province.

▌棕尾虹雉

拍摄地：西藏自治区

摄影：徐永春

Himalayan Monal

Lophophorus impejanus

Location: Tibet Autonomous Region

Photographer: Xu Yongchun

国家一级保护野生动物

羽色绚丽的棕尾虹雉栖息在自然条件严酷的高海拔地区。春夏白雾缭绕，它仍像一道高原彩虹飞翔在山谷间。

State First-class Protected Wild Animal

The Himalyan monal with gorgeous feathers inhabits harsh high-altitude areas. Shrouded in white mist in spring and summer, the flying monal looks like a rainbow over the valley.

绿尾虹雉

拍摄地：四川省
摄影：何晓安
Chinese Monal
Lophophorus lhuysii
Location: Sichuan Province
Photographer: He Xiaoan

国家一级保护野生动物

绿尾虹雉在高海拔陡峭的裸岩停歇，体羽在不同光线下变幻出多种色彩，人们赞誉它是秘境“青鸾”。“青鸾”在中国古典《山海经》中是吉祥美好的象征。

State First-class Protected Wild Animal

The Chinese monal rests on the steep bare rock at high altitude, and the color of its feathers varies under different lights. It is praised as the “Qingluan” in the secret land. “Qingluan” is a symbol of fortune and beauty in the Chinese literature of *The Classic of Mountains and Seas*.

白冠长尾雉

拍摄地：河南省
摄影：马晓光
Reeves’s Pheasant
Syrmaticus reevesii
Location: Henan Province
Photographer: Ma Xiaoguang

国家一级保护野生动物

白冠长尾雉超长的尾羽在求偶时高高竖起，更显妖娆。中国戏曲的道具“雉翎”取其高贵之意，作为身份象征。

State First-class Protected Wild Animal

The super long tail feathers of the Reeves’s pheasant look even more attractive when they are raised high in courting. The prop of Chinese opera “pheasant plume” takes its noble meaning as an identity symbol.

▌红腹锦鸡

拍摄地：河南省

摄影：徐永春

Golden Pheasant

Chrysolophus pictus

Location: Henan Province

Photographer: Xu Yongchun

国家二级保护野生动物

雪中奔跑的红腹锦鸡像一束风动的火焰，更像一幅现实版的凤凰齐飞。

State Second-class Protected Wild Animal

The golden pheasant running in the snow is like a bunch of blowing flame, even more like a realistic version of flying phoenixes.

▌高原山鹑
拍摄地：四川省
摄影：王治国
Tibetan Partridge
Perdix hodgsoniae
Location: Sichuan Province
Photographer: Wang Zhiguo

川西康定甲根坝，寒冷季节，栖息于高山灌丛地带的高原山鹑聚集在低洼处抱团取暖。

In the cold season in Jiagenba, Kangding, western Sichuan, Tibetan partridges that inhabit alpine shrubs gather in low-lying areas for warmth.

▌褐马鸡

拍摄地：山西省
摄影：李学良
Brown-eared Pheasant
Crossoptilon mantchuricum
Location: Shanxi Province
Photographer: Li Xueliang

国家一级保护野生动物

褐马鸡是我国特有雉类，仅见于中国山西、河北、陕西和北京的东灵山，性勇善斗，古人称之为勇雉。

State First-class Protected Wild Animal

The brown-eared pheasant, an endemic pheasant to China, is only found in Shanxi, Hebei, and Shaanxi, and Dongling Mountain of Beijing. Due to its nature of bravery and good at fighting, the brown-eared pheasant was called "brave pheasant" by the ancients.

血雉

拍摄地：四川省

国家二级保护野生动物

四川雅江帕姆岭大雪中的血雉。

黑嘴松鸡

拍摄地：内蒙古自治区

摄影：王治国

Black-billed Capercaillie

Tetrao urogalloides

Location: Inner Mongolia Autonomous Region

Photographer: Wang Zhiguo

国家一级保护野生动物

内蒙古根河，站在雪枝上求偶炫耀的黑嘴松鸡。

State First-class Protected Wild Animal

A showing off and courting black-billed capercaillie stands on a snowy branch in Genhe, Inner Mongolia.

绿孔雀

拍摄地：云南省
摄影：奚志农
Green Peafowl
Pavo muticus
Location: Yunnan Province
Photographer: Xi Zhinong

国家一级保护野生动物
春回大地，云南省中部元江畔的热带季雨林里，一只羽色艳丽的雄性绿孔雀漫步河滩。绿孔雀在中国仅分布于云南省境内，种群数量 555 ~ 600 只，属于极度濒危物种。

State First-class Protected Wild Animal

A colorful male green peafowl with gorgeous feathers strolls along a beach at the Yuanjiang River in a tropical monsoon rain forest in central Yunnan Province in spring.

The green peafowl is only distributed in Yunnan Province, with a population of 555—600, as a critically endangered species.

花冠皱盔犀鸟

拍摄地：云南省

摄影：郑山河

Wreathed Hornbill

Rhyticeros undulatus

Location: Yunnan Province

Photographer: Zheng Shanhe

国家一级保护野生动物

作为热带雨林旗舰物种，犀鸟是一片雨林生态健康的标志。云南盈江犀鸟谷已观测到4种犀鸟。

State First-class Protected Wild Animal

As the flagship species of the tropical rainforest, the presence of hornbills is a symbol of the ecologically healthy rainforest.

Four species of hornbills have been observed in the Hornbill Valley of Yingjiang, Yunnan Province.

冠斑犀鸟

拍摄地：云南省

摄影：丁彩霞

Oriental Pied Hornbill

Anthracoceros albirostris

Location: Yunnan Province

Photographer: Ding Caixia

国家一级保护野生动物

喂食后回望巢室的冠斑犀鸟。

冠斑犀鸟雌鸟在封闭的巢中孵卵，以避天敌。雄鸟每日觅食，无微不至地照顾雌鸟，因为它奇特的繁殖习性，当地少数民族亲切地称它为“爱情鸟”。

State First-class Protected Wild Animal

The oriental pied hornbill looking back at the nest after feeding.

The female hornbill hatches her eggs in a closed nest to avoid predators. The male forages every day and takes good care of the female. Local minorities affectionately call it “love bird” because of its unique reproductive behavior.

大鸨

拍摄地：内蒙古自治区
摄影：刘晶敏

Great Bustard
Otis tarda
Location: Inner Mongolia Autonomous Region
Photographer: Liu Jingmin

国家一级保护野生动物

每年 3 月，熬过了严冬的大鸨开始聚群，雄性大鸨昂首挺胸，翘起翅膀和尾羽，互相撞胸炫耀争雄，踱步回转，舞姿曼妙。有时这样的求偶动作会持续几个小时。

State First-class Protected Wild Animal

Every March, the great bustards begin to gather after the harsh winter. The males dance with heads, wings and tail feathers up. They collide chest and pace. Sometimes this courtship can last for hours.

普通楼燕

拍摄地：北京市

摄影：梁勤

Common Swift

Apus apus

Location: Beijing

Photographer: Liang Qin

1870 年在北京首次发现，也称“北京雨燕”。北京雨燕主要栖息在古建筑物或仿古建筑横梁的缝隙中，人们也习惯在忙碌中看到它们穿行于亭台楼阁的身影。环志研究发现，北京雨燕秋季最远迁徙到非洲越冬。

The common swift was first discovered in Beijing in 1870, hence it was named “Beijing swift” in Chinese. Common swifts mainly live in the gaps between the beams of ancient buildings. They can be seen flying through the pavilions. The bird banding study found that common swifts could migrate farthest to Africa for wintering.

红嘴相思鸟

拍摄地：重庆市

摄影：赵建英

Red-billed Leiothrix

Leiothrix lutea

Location: Chongqing

Photographer: Zhao Jianying

国家二级保护野生动物

羽色艳丽、鸣声婉转动听，在中国分布较广。但由于种群数量显著减少，已被列为国家二级保护野生动物加以保护。

State Second-class Protected Wild Animal

The red-billed leiothrix is widely distributed in China with gorgeous feathers and melodious song. However, due to significant population reduction, the red-billed leiothrix has been listed as State Second-class Protected Wild Animal.

台湾蓝鹊

拍摄地：台湾省
摄影：王艳秋
Taiwan Blue Magpie
Urocissa caerulea
Location: Taiwan Province
Photographer: Wang Yanqiu

台湾蓝鹊是台湾特有种，俗称“长尾山娘”，在低海拔山地，时常可见台湾蓝鹊的美丽身影。

The Taiwan blue magpie is endemic to Taiwan Province, which can be often seen in low-altitude mountains.

黑鸢

拍摄地：澳门特别行政区
摄影：许俊民
Black Kite
Milvus migrans
Location: Macao Special Administrative Region
Photographer: Xu Junmin

国家二级保护野生动物

在澳门城市上空盘旋的黑鸢。澳门拥有丰富的鸟类资源，也是候鸟的迁徙驿站。

State Second-class Protected Wild Animal

A black kite hovering over Macao, where rich in is bird resources and also serves as a stopover site for migratory birds.

雪鸮

拍摄地：内蒙古自治区
摄影：徐永春
Snowy Owl
Bubo scandiacus
Location: Inner Mongolia Autonomous Region
Photographer: Xu Yongchun

国家二级保护野生动物

雪鸮栖息于冻土和苔原地带，也见于荒地丘陵。以鼠类、鸟类、昆虫为食。在北极和西伯利亚繁殖，越冬时可见于中国北方部分地区，十分罕见。

State Second-class Protected Wild Animal

Snow owls inhabit permafrost and tundra regions and could also be found in barren lands and hills. They feed on rodents, birds and insects. Snow owls breed in Arctic and Siberia and can be rarely seen in parts of northern China during wintering.

高山兀鹫

拍摄地：青海省

摄影：徐永春

Himalayan Vulture

Gyps himalayensis

Location: Qinghai Province

Photographer: Xu Yongchun

国家二级保护野生动物

祁连山上空，天蓝如洗，太阳已经高高升起，半个月亮却还挂在天上。高山兀鹫在天穹盘旋，远远望去，越飞越高，似乎飞到了月亮之上。

State Second-class Protected Wild Animal

The sky over the Qilian Mountains is super blue. The sun has risen high while half of the moon is still hanging in the sky. Viewed from afar, the Himalayan vulture hovers in the sky and flew higher and higher close to the moon.

燕隼

拍摄地：北京市

摄影：关鹏

Eurasian Hobby

Falco subbuteo

Location: Beijing

Photographer: Guan Peng

国家二级保护野生动物

黄昏中，北京城区灯光初映，一只燕隼捕到一只蝙蝠。随着城市生态环境向好，燕隼、红隼等小型猛禽常在市区楼宇间筑巢繁衍。

State Second-class Protected Wild Animal

A Eurasian hobby caught a bat at dusk in downtown Beijing. With the improvement of urban ecological environment, small raptors such as Eurasian hobbies and common kestrels often nest and breed in urban buildings.

扬子鳄

拍摄地：安徽省

摄影：谢建国

Chinese Alligator

Alligator sinensis

Location: Anhui Province

Photographer: Xie Jianguo

国家一级保护野生动物

扬子鳄是中国特有的爬行动物，是世界上最小的鳄鱼之一，与恐龙属同一时代，历经几次“大灭绝”而奇迹般繁衍至今，有“活化石”之称。

State First-class Protected Wild Animal

The Chinese alligator, one of the smallest alligators in the world, is a reptile endemic to China. The ancestor of Chinese alligator is contemporaneous with dinosaurs. After several "mass extinction", the Chinese alligator has miraculously survived up to now, known as the "living fossil".

两栖、爬行动物

AMPHIBIANS AND REPTILES

爬行动物是真正适应陆栖生活的变温脊椎动物，与人类的关系密切。爬行动物体被鳞片或硬甲，在陆地繁殖。爬行类在中生代曾经盛极一时，种类繁多。中国现有爬行动物 511 种，包括蜥蜴和蛇、龟鳖和扬子鳄（*Alligator sinensis*）。

两栖动物是由水生到陆生的过渡类群，是人类熟知的一类野生动物，除南极洲和海洋性岛屿外，遍布全球。两栖动物对环境变化敏感，是生态环境健康的指示物种。中国共记录现生本土两栖动物 515 种，主要分布于秦岭以南，华西和西南山区属种最多。北京分布有黑斑侧褶蛙（*Pelophylax nigromaculatus*）、大蟾蜍（*Bufo Bufo*）等。

Reptiles are heterothermic vertebrates adaptive to terrestrial habitats. Reptiles have a covering of a special skin made up of scales or bony plates, and they breed on land. Reptiles were the dominant animals during the Mesozoic era. There are 511 species of reptiles in China, including lizards, snakes, turtles, tortoises and Chinese alligator (*Alligator sinensis*).

Amphibians are a group of organisms which represent the transition from aquatic life to terrestrial life. They are familiar to many people, and globally distributed except Antarctica and some oceanic islands. Amphibians are sensitive to environmental change, thus are an indicator of healthy ecological environment. There are 515 species of native amphibians recorded in China, and occurring in south of Qin Mountain, especially west and southwest of China. Black-spotted pond frog (*Pelophylax nigromaculatus*) and toad (*Bufo Bufo*) are often seen in Beijing.

扬子鳄

拍摄地：安徽省
摄影：谢建国
Chinese Alligator
Alligator sinensis
Location: Anhui Province
Photographer: Xie Jianguo

国家一级保护野生动物

野外放归的扬子鳄。

20 世纪末，由于生存环境恶化，野生扬子鳄数量急剧下降，已不足 150 条。1982 年成立安徽扬子鳄国家级自然保护区，至今已培育扬子鳄 15000 余条，累计野外放归 1000 余条。

State First-class Protected Wild Animal

A Chinese alligator released into the wild.

Due to the habitat degradation, at the end of the 20th century, the wild population of Chinese Alligator dropped significantly to less than 150 individuals. Anhui Chinese Alligator National Nature Reserve was established in 1982, and until 2021, more than 15,000 Chinese alligators have been bred and over 1,000 have been released into the wild.

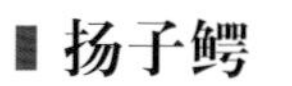

扬子鳄

拍摄地：浙江省
摄影：王聿凡
Chinese Alligator
Alligator sinensis
Location: Zhejiang Province
Photographer: Wang Yufan

国家一级保护野生动物

浙江湖州，雌性扬子鳄用泥土和树叶堆积起一个小土丘做巢。经过漫长的孵化期，小扬子鳄们从巢中探出头发出叫声，扬子鳄母亲听到叫声会来到巢边用嘴把扬子鳄带入水中，并会驱赶一切前来的入侵者。

State First-class Protected Wild Animal

A female Chinese alligator builds a nest with mud and leaves on a mound in Huzhou, Zhejiang Province. After a long incubation period, baby Chinese alligators poke their heads out of the nest and make calls. The mother alligator then would come to the nest and lead the youngs into the water by putting them in her mouth and chase away any intruders.

鳄蜥

拍摄地：广西壮族自治区
摄影：王聿凡
Chinese Crocodile Lizard
Shinisaurus crocodilurus
Location: Guangxi Zhuang Autonomous Region
Photographer: Wang Yufan

国家一级保护野生动物

溪流是鳄蜥的主要栖息地，白天停歇在水塘上方树枝上，一有风吹草动便纵身跳入水中逃之夭夭；夜晚在栖枝上睡觉。

State First-class Protected Wild Animal

The streams are the main habitat of the Chinese crocodile lizard. It hides and rests on perch over stream and escapes by jumping into water once disturbed during the day, and sleeps on branches.

中国大鲵

拍摄地：浙江省
摄影：周佳俊
Chinese Giant Salamander
Andrias davidianus
Location: Zhejiang Province
Photographer: Zhou Jiajun

国家一级保护野生动物

浙江青田，中国大鲵是现存世界上最大的两栖类，它们藏身于清澈的溪流或溶洞暗河之中，夜晚游出洞穴捕猎。

State First-class Protected Wild Animal

The Chinese giant salamander, the world’s biggest living amphibian, hides in clear streams and caves and emerges at night to prey in Qingtian, Zhejiang Province.

秉志肥螈（幼体）

拍摄地：浙江省
摄影：王聿凡
Pingchi's Newt
Pachytriton granulosus
Location: Zhejiang Province
Photographer: Wang Yufan

秉志肥螈的取名是为了纪念中国动物学奠基人秉志先生，这种有尾目动物的成体在我国东部的高山溪流中并不少见，但拍到幼体的照片就十分难得了。秉志肥螈的幼体有三对羽毛状外鳃非常可爱。

The Pingchi's newt was named in memory of Mr. Bingzhi, the founder of Chinese zoology. The adult newts are common in mountain streams in eastern China, but photographs of the larvae are rare. The Pingchi's newt's larvae has three lovely pairs of feathery outer gills.

安吉小鲵
拍摄地：浙江省
摄影：王聿凡
Anji Hynobiid
Hynobius amjiensis
Location: Zhejiang Province
Photographer: Wang Yufan

国家一级保护野生动物
浙江湖州，安吉小鲵的亚成体身上密布浅蓝色的碎斑，如同星河，随着年龄增长这些斑纹将慢慢褪去。安吉小鲵仅分布在天目山脉安吉龙王山保护区的几个高山沼泽之中，高山沼泽实为死火山口。

State First-class Protected Wild Animal

The sub-adult of Anji hynobiid in Huzhou, Zhejiang Province is covered with light blue spots like a galaxy, which will slowly fade with age. Anji hynobiids are only distributed among several alpine swamps that are actually in the dead volcanic craters in the Anji Longwang Mountain Nature Reserve in the Tianmu Mountains.

镇海棘螈

拍摄地：浙江省

摄影：王聿凡

Chinhai Spiny Newt

Echinotriton chinhaiensis

Location: Zhejiang Province

Photographer: Wang Yufan

国家一级保护野生动物

镇海棘螈仅分布在浙江省宁波市，它虽然是两栖类，但成年的镇海棘螈却不能再下水，就算产卵也是在水塘边的落叶堆下，等到下雨或涨水，发育的卵才会进入水塘之中。

State First-class Protected Wild Animal

The Chinhai spiny newt is only distributed in Ningbo City, Zhejiang Province. Although it is an amphibian, the adult newt can no longer live in the water. Eggs are laid in the fallen leaves near the pond. The developed eggs will enter the pond while raining or the water rises.

黑疣大壁虎

拍摄地：香港特别行政区
摄影：吴颖
Reeves' Tokay Gecko
Gekko reevesii
Location: Hong Kong Special Administrative Region
Photographer: Wu Ying

国家二级保护野生动物

黑疣大壁虎是近年从大壁虎（*Gekko gecko*）分出来的独立物种，是世界上最大的壁虎之一，体长可达 30 多厘米。它们白天通常躲藏在岩石裂缝、房屋缝隙中，夜间出来活动。除了昆虫等节肢动物外，它们还会捕食其他小型的壁虎，甚至是小鸟。

State Second-class Protected Wild Animal

The Reeves' tokay gecko is a species recently independent from tokay gecko (*Gekko reevesii*). It is one of the biggest geckos in the world with a body length of more than 30 centimetres. They usually hide in cracks of rocks and houses during the daytime and come out at night. Besides insects and other arthropods, they also prey on other small geckos and even small birds.

西藏温泉蛇

拍摄地：西藏自治区

摄影：王聿凡

Tibetan Hot-spring Snake

Thermophis baileyi

Location: Tibet Autonomous Region

Photographer: Wang Yufan

国家一级保护野生动物

西藏温泉蛇是世界上分布海拔最高的蛇类，它们生活的地方每年有半年时间是冻土，为了不让自己被冻住，它们选择在温泉泉眼附近度过漫长的冰封期。

State First-class Protected Wild Animal

The Tibetan hot-spring snake is the species with the highest-altitude distribution in the world. The place where they live is frozen soil for half a year each year. In order to prevent themselves from being frozen, they choose to spend a long period of freezing near the hot spring.

角原矛头蝮

拍摄地：浙江省
摄影：王聿凡
Horned Pitviper
Protobothrops cornutus
Location: Zhejiang Province
Photographer: Wang Yufan

国家二级保护野生动物

角原矛头蝮的眼球上方有两片鳞极度延伸，形成了角状的凸起，据推测是有利于伪装。角原矛头蝮非常依赖火山岩或石灰岩的环境。

State Second-class Protected Wild Animal

A hornlike bulge, which is formed by two extremely elongated scales above the horned pitviper's eyeballs, maybe is beneficial to camouflage. Horned pitvipers are highly dependent on volcanic or limestone environments.

尖吻蝮

拍摄地：浙江省
摄影：王聿凡
Sharp-snouted Pitviper
Deinagkistrodon acutus
Location: Zhejiang Province
Photographer: Wang Yufan

因为吻端有一个突起结构顾名为尖吻蝮，它有个更为人所知的名字叫“五步蛇”。

It is called “sharp-snouted pitviper” in Chinese because of a protuberant structure at the end of its snout. It has a better known name as the “five-pace snake”.

平胸龟

拍摄地：云南省

摄影：曾祥乐

Big-headed Turtle

Platysternon megacephalum

Location: Yunnan Province

Photographer: Zeng Xiangle

国家二级保护野生动物

平胸龟又叫鹰嘴龟，是长相最奇特的龟。巨大的头部，与体长相近的尾巴，灵活的四肢，不仅可以在溪流中追逐螃蟹，甚至还能上树，一改人们对龟类笨拙体态的认知。

State Second-class Protected Wild Animal

The big-headed turtle called as "eagle-billed turtle" is the strangest looking turtle. The big head, with a tail in the same length of body and flexible limbs allow it to chase crabs through streams, and even climb up trees, completely altering the perceptions of clumsiness of turtles.

斑鳖

拍摄地：江苏省

摄影：齐硕

Yangtze Giant Softshell Turtle

Refetus swinhoei

Location: Jiangsu Province

Photographer: Qi Shuo

国家一级保护野生动物

斑鳖是世界上最大的鳖之一，曾经广泛分布在中国长江中下游至越南。如今已知个体仅剩3只，国内仅1只存活在苏州上方山森林动物园。

State First-class Protected Wild Animal

The Yangtze giant softshell turtle is one of the largest softshell turtles in the world, once widely distributed in the middle and downstreams of the Yangtze River, China and Vietnam. Now there are only three known individuals, and only one is raised in Suzhou Shangfang Mountain Forest Zoo in China.

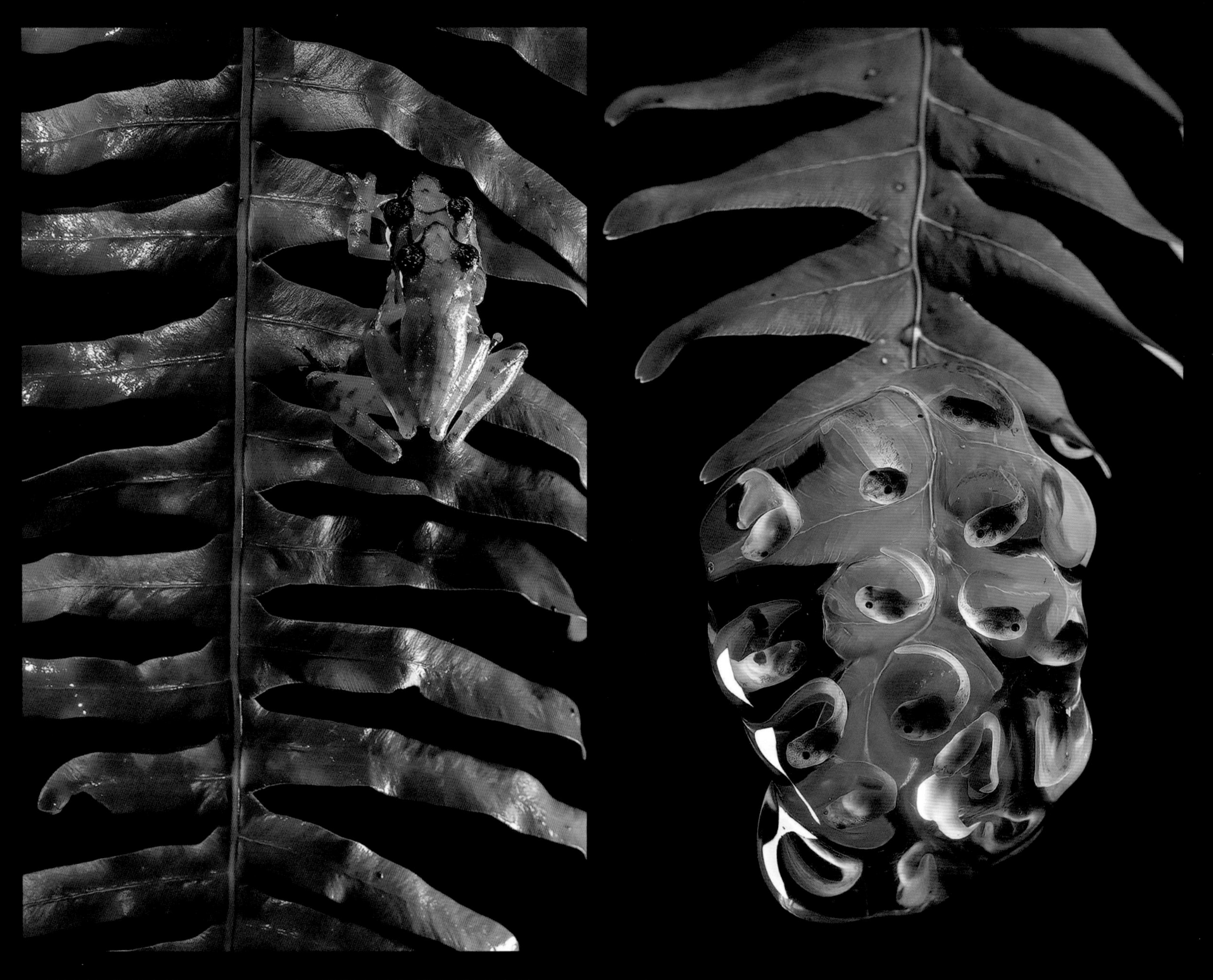

黑眼睑纤树蛙

拍摄地：广东省
摄影：王聿凡
Black Eye-lidded Small Treefrog
Gracixalus gracilipes
Location: Guangdong Province
Photographer: Wang Yufan

黑眼睑纤树蛙生活的地区降水充沛，蕨类叶子为了快速排水会把雨水汇集到叶尖，树蛙便把卵产在水潭上方的树叶尖，等卵发育成蝌蚪就可以落入下方水中。一来可以躲避水中的鱼类捕食蛙卵，二来能保存水分。

The habitat of black eye-lidded small treefrogs is abundant in rainfall. The tree frog lays its eggs on the tips of fern leaf above the pool. When it rains, the structure of fern leaf would collect rainwater to the left tip for rapid drainage. The hatched tadpoles could fall into the pool along with rainwater to avoid fish predation, as welll as maintaining the moisture of egges.

峨眉髭蟾

拍摄地：四川省

摄影：王聿凡

Emei Moustache Toad

Vibrissaphora boringii

Location: Sichuan Province

Photographer: Wang Yufan

国家二级保护野生动物

峨眉髭蟾雄性在繁殖期的时候，上唇会长出一圈角质刺，用角质刺充当武器来驱赶其他雄性争抢地盘。

State Second-class Protected Wild Animal

During the breeding season, male Emei moustache toads grow a ring of horny spines on their upper lip, which are used as weapons to ward off other males and compete for territory.

▮ 阳彩臂金龟

拍摄地：浙江省

摄影：王聿凡

Caiyang Long-armed Beetle

Cheirotonus jansoni

Location: Zhejiang Province

Photographer: Wang Yufan

国家二级保护野生动物

阳彩臂金龟是我国体重最大的昆虫之一，雄性有着不合比例的“长臂”，上面还有夸张的尖刺，这是它们为了争夺领地演化出的武器。阳彩臂金龟的幼虫取食被真菌分解后的朽木，因此它们非常依赖健康的常绿阔叶林生境。

State Second-class Protected Wild Animal

The Caiyang long-armed beetle is one of the biggest insects in China. The males have disproportionate long arms with exaggerated spikes which are evolved as weapons to fight for territory. The larvae feed on decayed wood decomposed by fungi, and therefore they are highly dependent on a healthy evergreen broad-leaved forest habitat.

昆虫

INSECTS

昆虫种类繁多、形态各异，属于无脊椎动物中的节肢动物，是地球上数量最多的动物群体。最近的研究表明，全世界的昆虫可能有 1000 万种，约占地球所有生物物种的一半，踪迹几乎遍布世界的每一个角落。

中国的昆虫种类占世界种类的 1/10，75 种昆虫列入《国家重点保护野生动物名录》，其中，金斑喙凤蝶（*Teinopalpus aureus*）、中华蛩蠊（*Galloisiana sinensis*）、陈氏西蛩蠊（*Grylloblattella cheni*）为国家一级保护野生动物。

Insects are a remarkably diverse group of animals including in morphology. All insects belong to the phylum Arthropoda in invertebrates. Insects are the most numerous among animals on the globe. It is estimated that there are currently ten million species of insects, almost half of all the wildlife, and they live in all habitats.

Ten percent of insect species in the world occur in China, with seventy-five species listed in the *State Key Protected Wild Animal List*, including several are listed as State First-class Protected Wild Animal, e.g golden kaiserihind (*Teinopalpus aureus*), ice crawlers (*Galloisiana Sinensis*), *Grylloblattella cheni*.

兴安蝴蝶泉
拍摄地：内蒙古自治区
摄影：郭启明
Xingan Butterfly Spring
Location: Inner Mongolia Autonomous Region
Photographer: Guo Qiming

内蒙古大兴安岭阿鲁省级自然保护区

Alu Provincial Nature Reserve, Great Khingan Mountains, Inner Mongolia

金斑喙凤蝶

拍摄地：江西省
摄影：陈敢清
Golden Kaiserihind
Teinopalpus aureus
Location: Jiangxi Province
Photographer: Chen Ganqing

国家一级保护野生动物

江西省九连山国家级自然保护区。金斑喙凤蝶被誉为“蝶之骄子”，珍贵而稀少，是中国唯一的蝶类国家一级保护野生动物，排在世界八大名贵蝴蝶之首。

State First-class Protected Wild Animal

Photographed in Jiulianshan National Nature Reserve of Jiangxi Province. The golden kaiserihind is known as “the pride of butterflies”. Because of the rarity and preciousness, it is listed as State First-class Protected Wild Animal, the only protected butterfly in China, and ranks the first among the top eight precious butterflies in the world.

君主绢蝶

拍摄地：青海省
摄影：黄亚慧
Emperor Apollo
Parnassius imperator
Location: Qinghai Province
Photographer: Huang Yahui

国家二级保护野生动物

祁连山。大型绢蝶，翅白色，前翅发育灰色和黑色斑条，后翅中域有两枚圆形红斑，亚缘有一条灰带，臀角一般有两枚卵形蓝斑，多见于 1500 ~ 5000 米的山区灌草地、石坡，一年一代，成虫多见于 7 月。

State Second-class Protected Wild Animal

Photographed in Qilian Mountains. The emperor apollo is a big size snow apollos with white wings, gray and black stripes on the fore-wings, and two circular red spots in the middle of the hind-wings, a gray band on the submargin, and two oval blue spots on the inner angle. It's mostly seen in shrub grassland and stone slope in mountainous areas of 1500 – 5000 meters, has one generation per year, and adults are mostly found in July.

中华虎凤蝶

拍摄地：湖南省
摄影：张京明
Chinese Luehdorfia
Luehdorfia chinensis
Location: Hunan Province
Photographer: Zhang Jingming

国家二级保护野生动物

中华虎凤蝶是中国独有的一种野生蝶，被昆虫专家誉为“国宝”。

乌云界国家级自然保护区，是我国中华虎凤蝶种群数量最大的地区之一。近年，该保护区加强对中华虎凤蝶栖息地保护，种群数量稳定在1000多只。

State Second-class Protected Wild Animal

The Chinese luehdorfia is an endemic species to China, which is praised as “national treasure” by entomologists.

Wuyunjie National Nature Reserve is one of the areas with the largest population of Chinese luehdorfia in China. In recent years, the reserve has strengthened the habitat conservation of the Chinese luehdorfia, and the population has stabilized at more than 1,000 individuals.

兰花螳螂
拍摄地：云南省
摄影：罗爱东
Orchid Mantis
Hymenopus coronatus
Location: Yunnan Province
Photographer: Luo Aidong

兰花螳螂是螳螂中最漂亮的物种之一。它们的步肢演化出类似花瓣的构造和颜色，可以拟态成一朵花而不被猎物察觉。

Orchid mantis is one of the most beautiful species among mantises, its legs have evolved into petal-like structure and color, allowing them to mimic a flower without being detected by prey.

黄猄蚁

拍摄地：广西壮族自治区
摄影：王聿凡
Weaver Ant
Oecophylla smaragdina
Location: Guangxi Zhuang Autonomous Region
Photographer: Wang Yufan

广西崇左，黄猄蚁和叶蝉有着奇妙的合作关系，黄猄蚁会用触角轻轻敲击叶蝉若虫的身体，叶蝉便会领会其意，分泌一滴小小的蜜露；黄猄蚁便会收集这些蜜露作为食物，而叶蝉则能得到黄猄蚁全天候的保护。

Weaver ants and leaf hoppers (Cicadellidea) have a wonderful cooperative relationship. Ants tap the body of leaf hoppers gently with their antennae, and leaf hoppers will secrete a small drop of honeydew. Weaver ants can collect honeydew for food, and leaf hoppers can get protection from ants.

萤火虫

拍摄地：江苏省

摄影：刘晓勤

Firefly

Lampyridae

Location: Jiangsu Province

Photographer: Liu Xiaoqin

萤火虫是最为人熟知的发光动物，每当发生季的夜晚漫天萤火。不同的萤火虫有着不同的发光频率，用长曝光和堆栈的拍摄方式可以留下萤火虫发光飞行的轨迹。

Fireflies are the most well-known luminescent animals, and fill the sky at night during the mating season. Different species of fireflies have different luminous frequencies, and we can record the luminous flight path of fireflies with long exposure and stack shooting.

长江江豚
拍摄地：江西省
摄影：王筱华
Yangtze Finless Porpoise
Neophocaena asiaeorientalis
Location: Jiangxi Province
Photographer: Wang Xiaohua

国家一级保护野生动物

两只江豚同时跃出水面，欲比高低。

位于南昌市东湖区扬子洲镇境内的赣江水域常有江豚成群捕食竞游的场景。

State First-class Protected Wild Animal

Two Yangtze finless porpoises are leaping out of the water together.

The scene of porpoises hunting and swimming in groups often can be seen in the Ganjiang River, located in Yangzizhou Town, Donghu District, Nanchang City.

水生动物

AQUATIC ANIMALS

水生动物是指主要在水中生活的动物，按照栖息场所可分为海洋动物和淡水动物两种。在中国辽阔的土地上，江河纵横交错，湖泊星罗棋布，内陆水体面积约占全国总面积的 1/50，水生动物资源异常丰富，特别是中国特有的淡水动物种类很多，仅鱼类就有 400 种。

中国的海洋动物种类约占全世界海洋动物总种数的 10%。鱼类、头足类和虾、蟹类是最主要的海洋动物。海洋动物种数的分布趋势是南多北少，即南海的种类较多，而黄海、渤海的种类较少。

水生哺乳动物是水生动物中重要的类群。中国有 50 余种，绝大部分为受保护动物。水生哺乳动物中大部分是鲸类动物，中国鲸类动物均为国家重点保护野生动物，其中白鱀豚 (*Lipotes vexillife*)、中华白海豚 (*Sousa chinensis*)、长江江豚 (*Neophocaena asiaeorientalis*) 为国家一级保护野生动物。

Aquatic animals are those animals that live all the time or much of their life in the water divided into freshwater or marine species. The rivers and lakes in China are numerous covering 2% of the terrestrial area. Aquatic animal resources are extremely abundant in China, especially the freshwater species, with over 400 fish species.

Ten percent of all ocean species occur in Chinese seas, (fish, cephalopods, crustaceans). There are more species in southern waters (e.g., the South Sea) than northern (e.g., the Yellow Sea, Bo Sea).

Aquatic mammals are important groups of aquatic animals. There are more than 50 species in China, most of which are protected animals. Most aquatic mammals are cetaceans. All cetaceans are listed as the State Key Protected Wild Animal, among which the baiji (*Lipotes vexillifer*), Chinese white dolphin (*Sousa chinensis*) and Yangtze finless porpoise (*Neophocaena asiaeorientalis*) are listed as State First-class Protected Wild Animal.

中华白海豚

拍摄地：广东省
摄影：冯抗抗
China White Dolphin
Sousa chinensis
Location: Guangdong Province
Photographer: Feng Kangkang

国家一级保护野生动物

渔民夫妻俩在专心地收着网，船边突然跃起一头中华白海豚，从它的肤色来看应该比较年轻。

State First-class Protected Wild Animal

When the couple are intently pulling their net, a Chinese white dolphin jumps up beside the boat. It should be relatively young individual judging from its skin color.

中华白海豚

拍摄地 : 广东省
摄影 : 冯抗抗
Chinese White Dolphin
Sousa chinensis
Location: Guangdong Province
Photographer: Feng Kangkang

国家一级保护野生动物

正在捕食的中华白海豚。

中华白海豚是哺乳类动物，在中国主要分布于东南部沿海。以河口的咸淡水鱼类为食。虽名为“白海豚”，但刚出生的中华白海豚呈深灰色，年青的呈灰色，成年的则呈粉红色。

State First-class Protected Wild Animal

A hunting Chinese white dolphin.

The Chinese white dolphin is a mammal, which is mainly distributed in the southeast coast of China. It feeds on saltwater and freshwater fishes in the estuary. Although named “white dolphin” in Chinese, the color lightens with age. The newly born dolphin is in dark gray, sub-adult in gray, and the adult turns pink.

长江江豚

拍摄地：湖北省

摄影：武明录

Yangtze Finless Porpoise

Neophocaena asiaeorientalis

Location: Hubei Province

Photographer: Wu Minglu

国家一级保护野生动物

湖北宜昌葛洲坝库区下游，一只江豚跃出逐浪嬉戏。

江豚形态可爱，被称为长江的微笑。2017 年调查，野生江豚仅存 1012 只，目前已建立 8 个江豚自然保护区。中国 2020 年开始实施长江十年禁捕，中国的保护工作者希望通过新的禁渔和污染防治措施，江豚能够重新在中国最大的河流中畅游。

State First-class Protected Wild Animal

A Yangtze finless porpoise jumping out and playing in the downstream of Gezhouba Dam Reservoir in Yichang, Hubei Province.

The Yangtze Finless porpoise has an adorable shape and is called the smile of the Yangtze River. According to the survey in 2017, only 1,012 Yangtze finless porpoises remained in the wild. Eight Yangtze finless porpoise nature reserves have been established. China has been implemented a 10-year fishing ban in the Yangtze River since 2020. Conservationists in China are hoping that through the new fishing ban and pollution control measures, Yangtze finless porpoises could freely swim again in China's largest river.

布氏鲸

拍摄地：广西壮族自治区
摄影：陈默
Bryde's whale
Balaenoptera edeni
Location: Guangxi Zhuang Autonomous Region
Photographer: Chen Mo

国家一级保护野生动物

广西涠洲岛，布氏鲸捕鱼现场，成群的鸥鸟追随布氏鲸，分享这场盛宴。

2016 年以来，布氏鲸频频现身北部湾涠洲岛海域，为了保护布氏鲸，当地政府于 2018 年颁布了《北海市涠洲岛生态环境保护条例》。

State First-class Protected Wild Animal

A flock of seagulls follows Bryde's whale preying and share the feast in Weizhou Island, Guangxi Zhuang Autonomous Region .

Since 2016, Bryde's whales have frequently appeared in the waters of Weizhou Island in the Beibu Gulf. In order to protect Bryde's whales, the local government issued the *Regulation on the environmental conservation of Weizhou Island in Beihai City* in 2018.

西太平洋斑海豹

拍摄地：山东省

摄影：顾晓军

Spotted Seal

Phoca largha

Location: Shandong Province

Photographer: Gu Xiaojun

国家一级保护野生动物

山东长岛，西太平洋斑海豹喜欢在退潮的时候躺在岩石上晒太阳。这只西太平洋斑海豹打哈欠的样子如同掩面窃笑，十分可爱。

State First-class Protected Wild Animal

In Changdao, Shandong Province, the spotted seals enjoy basking on rocks at low tide .This spotted seal yawns like a adorable snicker.

西太平洋斑海豹

拍摄地：山东省

摄影：顾晓军

Spotted Seal

Phoca largha

Location: Shandong Province

Photographer: Gu Xiaojun

国家一级保护野生动物

西太平洋斑海豹是海洋性哺乳动物，在中国分布于渤海和黄海北部。辽东湾是它们重要的繁殖区。

State First-class Protected Wild Animal

Spotted seals are marine mammals, mainly distributed in the Bohai Sea and the northern Yellow Sea in China. Liaodong Bay is its important breeding site.

北海狮

拍摄地：山东省

摄影：王成军

Steller Sea Lion

Eumetopias jubatus

Location: Shandong Province

Photographer: Wang Chengjun

国家二级保护野生动物

2020 年 7 月 3 日，一只北海狮现身长岛北隍城岛，这也是长岛海域首次发现北海狮。

北海狮是海狮科最大的一种，分布于太平洋海域，在中国十分罕见。

State Second-class Protected Wild Animal

On 3 July 2020, a steller sea lion appeared at north Huangcheng Island of Changdao Island. This is the first steller sea lion recorded in Changdao Island.

The steller sea lion is the largest species of the Otariidae. It is commonly distributed in the Pacific Ocean but rarely seen in the seas of China.

绿海龟

拍摄地：台湾省

摄影：马格努斯·隆格伦

Green Turtle

Chelonia mydas

Location: Taiwan Province

Photographer: Magnus Lundgren

国家一级保护野生动物

绿海龟是中国海域最为常见的海龟，之所以叫绿海龟是因为它们的脂肪呈现绿色。塑料垃圾、渔网对它们有着巨大威胁。

State First-class Protected Wild Animal

Green turtles are the most commonly seen sea turtles in the seas of China. They are called green turtles because of their greenish-color fat. Plastic waste and fishing nets pose a serious threat to their survival.

▌花鹿角珊瑚

拍摄地：海南省

摄影：周佳俊

Branch Coral

Acropora florida

Location: Hainan Province

Photographer: Zhou Jiajun

国家二级保护野生动物

这是海底人工造礁，帮助已经被破坏的珊瑚礁快速恢复的一种方式。

State Second-class Protected Wild Animal

The construction of artifical reef is a measure for quick recovery of destroyed coral reefs.

黄海马

拍摄地：香港特别行政区
摄影：马格努斯・隆格伦
Great Seahorse
Hippocampus kelloggi
Location: Hong Kong Special Administrative Region
Photographer: Magnus Lundgren

国家二级保护野生动物

海马是唯一一种采用直立游泳方式的小型鱼类，体长可达17~30 厘米，其孵化由雄性在育儿袋中完成。

海马曾遭到大量捕捞，以至于面临生存危机。2004 年，我国把海马纳入国家二级保护野生动物。

State Second-class Protected Wild Animal

The seahorses are the only small fish that swim upright. The size can reach 17 to 30 centimeters in length, and baby seahorses are hatched in the males' pouch.

The seahorse has been overexploited, resulting in the crisis of survival. In 2004, China listed all seahorse species as State Second-class Protected Wild Animals.

松江鲈

拍摄地：浙江省
摄影：李帆
Roughskin Sculpin
Trachidermus fasciatus
Location: Zhejiang Province
Photographer: Li Fan

国家二级保护野生动物

“西风斜日鲈鱼香”，被古人赞不绝口的鲈鱼其实是杜父鱼科的松江鲈，因鳃盖膜上两道橘红色斜带酷似两片鳃，又名“四鳃鲈”。

State Second-class Protected Wild Animal

The ancients said that autumn is the season of sculpin.

The sculpin praised by the ancients due to the delicious taste is actually the roughskin sculpin. Because of the two orange stripes on the operculum membrane resemble two gills, it is also called “four-gilled sculpin”.

湟鱼

拍摄地：青海省

摄影： 星智

Przewalski's Naked Carp

Gymnocypris przewalskii

Location: Qinghai Province

Photographer: Xing Zhi

每年 6~7 月是湟鱼的繁殖季节，成群湟鱼沿着青海湖的支流中逆流而上，向着延续生命的产卵地进发，“半河清水半河鱼”的形容一点也不夸张。

June to July each year is the breeding season of Przewalski's naked carp and schools of Przewalski's naked carp go upstream along the tributaries of Qinghai Lake heading towards the spawning place. The scene is so spectacular that it is not an exaggeration to describe the river as half with clear water and half with fish.

唐鱼

拍摄地：广西壮族自治区
摄影：周佳俊
White Cloud Mountain Minnow
Tanichthys albonubes
Location: Guangxi Zhuang Autonomous Region
Photographer: Zhou Jiajun

国家二级保护野生动物

广西防城港，泥底溪流中，唐鱼穿梭在宽叶隐棒花丛中。唐鱼又叫“白云金丝鱼”，因美丽的外表也成为著名的原生观赏鱼。

State Second-class Protected Wild Animal

The white cloud mountain minnows shuttle among *Cryptocoryne aquarium* plants in a mud-bottom stream in Fangchenggang, Guangxi Zhuang Autonomous Region. It is also a famous native ornamental fish due to its beautiful appearance.

田林金线鲃

拍摄地：广西壮族自治区
摄影： 周佳俊
Sinocyclocheilus tianlinensis
Location: Guangxi Zhuang Autonomous Region
Photographer: Zhou Jiajun

国家二级保护野生动物

广西百色暗无天日的地下河，是田林金线鲃的家园，它无眼无鳞，极度特化适应这特殊的生存环境。

State Second-class Protected Wild Animal

The dark underground river is the home of *Sinocyclocheilus tianlinensis* in Baise County, Guangxi Zhuang Autonomous Region. Eyeless and scaleless apparence is an extremely specialized feature for the adaption of special living condition.

后记

大兴安岭林间健硕威猛的东北虎，秦岭山中温情育仔的大熊猫，西双版纳雨林中睡姿萌动的亚洲象，鄱阳湖湿地成群觅食嬉戏的白鹤，南海大襟岛海域游弋跳跃的中华白海豚，汉中洋县山野间栖息鸣唱的朱鹮，燕山大海坨主峰下比翼竞翔的天鹅……

当这本画册呈现在你面前的时候，相信你会被中国丰富多样的生物物种惊艳、震撼，这是生命的魅力，也是影像的力量！这些影像真实地记录下中国野生动物的精彩瞬间，展现出迷人的野性之美。这些影像记录说明，在中国，人与自然、人与野生动物的关系正在发生积极的变化，人与野生动物相互重新认识，中国野生动物保护取得了积极的成果。

《中国野生动物》是“自然影像中国”项目多年成果的集萃。画册选用 75 位中外摄影师的 150 余幅照片，涵盖全国 34 个省级行政区的 120 余种国家重点保护野生动物。为了拍摄这些珍贵的影像，摄影师付出了极大的艰辛。这些作品是他们与野生动物相处互信的成果和爱心的呈现。

最后，真诚地感谢为本画册提供图片的摄影师及付出智慧和辛劳的专家学者。

POSTSCRIPT

The strong and powerful tiger among forests in Great Khingan, the tender mothering giant panda in Qin Mountains, the sleeping Asian elephant among rain forests of Xishuangbanna, the foraging flocks of Siberian crane in Poyang Lake, swimming and leaping Chinese white dolphin in the South Sea, the singing crested ibis in the wild in Yang County, the swans flying at Dahaituo of Yan Mountain…

We hope you have been astonished and amazed by the rich range of species and their charm through reading and seeing the images in this book. These images record the moment and beauty of the wild. It shows the active connection between people and the rest of nature, the links between human and wildlife, and the continuing achievements in wildlife conservation.

This book is the jewel in the crown of the “Natural Image China” project with 150 images from 75 photographers in China and beyond. It covers 120 species of China’s key protected animals in all the 34 provinces, despite difficulties in taking the images through the ability and caring of the photographers, showing the mutual trust between people and wildlife.

We would like to express our appreciation to all the photographers and experts who have contributed to the success of this book.

轴孔珊瑚属的物种和长砗磲

拍摄地：海南省
摄影：周佳俊
Staghorn Coral and Small Giant Clam
Acropora sp. and Tridacna maxima
Location: Hainan Province
Photographer: Zhou Jiajun

均为国家二级保护野生动物

它们是南海珊瑚礁生态系统重要的组成生物。长砗磲是世界上最大的贝类之一，曾被大量捕捞做成贝壳饰品，近年的保护力度加大和人工繁育的成功，它们在南海数量逐渐恢复。

State Second-class Protected Wild Animal

Staghorn coral and small giant clam are important components of coral reef ecosystems in the South China Sea. Small giant clam is one of the largest tridacna species in the world. It has been extensively fished for shell ornaments. In recent years, with increased conservation efforts and the success of artificial breeding, its population is gradually recovering in the South China Sea.